Around Switzerland in 80 Maps

A magical journey

Diccon Bewes

For my grandmother, who loved maps
but was a terrible map-reader

Contents

Foreword

Maps are one of the oldest means of communication. Before people could write, read or count, they were aware of their everyday environment and drew sketches of it. But maps are also as modern as never before. In a society dominated by visual information, the number of new maps is increasing almost exponentially. Because maps bring things into focus. With their very own graphic language, they help to easily and efficiently illustrate the increasingly complex issues of the world. If "a picture is worth a thousand words" then "a map is worth a thousand facts".

Switzerland is a country of maps. For centuries all kinds of maps have been produced and distributed here. With their native drive for quality and perfection, Swiss engineers and cartographers produced extremely accurate maps and atlases, which became famous worldwide and have been imitated in many other countries. Fortunately for map makers, Switzerland has also always been a land of map readers. In perhaps no other country is "map reading" taught so intensively, from an early age at home, at school or later in the military. Maps accompany Mr and Mrs Schweizer through their whole lives and are part of their identity. While other countries simply have topographic maps, Switzerland has a "National Map", one that is a national cultural treasure, where even the slightest graphical changes – such as in the latest edition – are suspiciously eyed and scrutinized by thousands of users.

It's surprising that no one else ever came up with the idea of taking a popular visual journey through more than 500 years of the rich tradition of Swiss maps. Maybe it really did just need the celebrated "outsider perspective", as well as the carefree curiosity of an Englishman who has lived in and travelled around Switzerland for many years. Diccon Bewes has successfully tackled this undertaking and collected his "Best of" maps of and about Switzerland for us. In the style of Jules Verne, he trips jauntily through space and time in 80 maps, covering almost all regions of Switzerland, as well as the significant historical events of the last few centuries that have shaped our country. And I'm happy to say that the map always stands centre stage. As with a traditional atlas, you do not have to meticulously read this book from start to finish. You also do not have to be a historian or geographer to understand it. Instead you can comfortably leaf through the maps and stop at any one. With few words you can still get a very good idea of Switzerland's identity. The appearance and types of maps vary from page to page, so that you'll never get bored. And along the way, Diccon Bewes writes a short modern history of Swiss cartography, for which he is warmly thanked.

I am extraordinarily pleased that this book is being published at this point in time. That's because the International Cartographic Association, with the support of the United Nations, has declared 2015/16 to be 'International Year of the Map'; Switzerland is also participating. Just as language helps us to express ourselves without revealing everything, so maps allow us to understand important spatial structures and developments without having to travel everywhere.

This cultural achievement, which we use on a daily basis, often without thinking, should be brought back into the people's consciousness. Diccon Bewes' book is ideally suited for this. For, beyond the scientific considerations for the production and use of these maps, they are - as the author writes - "simply beautiful"! Nothing else need be said.

Dr Thomas Schulz

President of the Swiss Society of Cartography SSC

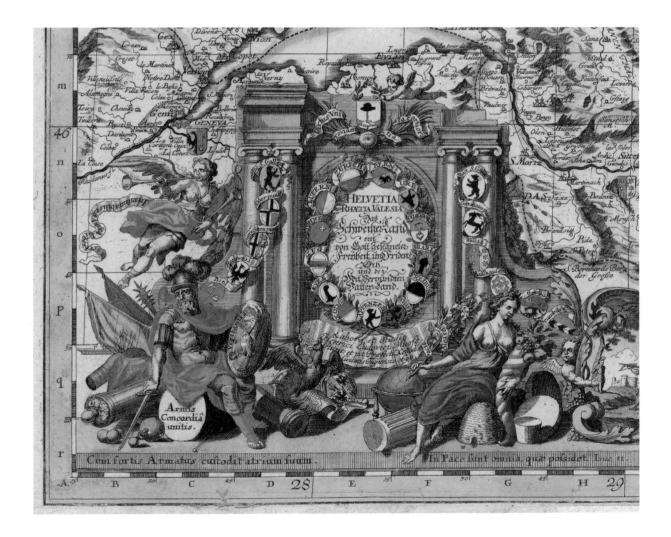

Introduction

A map can be decorative or informative, or both; it can reveal geography or history, or both; it can enlighten or deceive, or both. In fact, a map can do all of the above, all at the same time. It is the perfect pictorial way to explain and entertain. Together, the maps in this book achieve both while giving a detailed picture of Switzerland.

These 80 maps are contemporary to the eras they portray, rather than drawn for this book. The majority lie in archives in Switzerland, from the vast collection of Zurich Central Library to the small repositories of individual companies. Each has been chosen for a particular reason, be that an important moment in Swiss history or an interesting aspect of Swiss life. Or because it's simply beautiful.

There are no existing maps of Switzerland from its early years, when the legends of William Tell and the Rütli oath are set, so our story begins as the 15th century ends. Early maps were often drawn from the mapmaker's perspective, which in Switzerland usually meant sitting in Zurich or Bern and looking towards the Alps on the horizon. In other words, south at the top. We take for granted that modern maps are oriented to the north but, as this book will show, that isn't always the case.

Through these maps we chart the growth and development of Switzerland over the centuries. Its borders were shaped by conquest and defeat until finally being established in 1815, while the country's name has changed as often as its shape, thanks to both language and politics. As Latin disappeared so too did Helvetia on the maps, replaced by linguistic variations of Swissness.

But it wasn't always clear if Switzerland actually was one country. For many years it was more than a loose alliance but not quite a single entity, as the decorative cartouche here shows. The modern French name – *La Suisse* – is singular but on this map from 1703 it is *Les Suisses*, plural: "The Swiss, their allies and their subjects". Or almost like 'the Switzerlands', a confederation of 13 cantons represented here by their coats of arms.

Hand-painted or computer-generated, medieval or modern, city or country, all the maps have one thing in common: Switzerland. But this isn't a book with 80 maps of Switzerland, as many of them don't cover the whole country; it also isn't a book of 80 Swiss maps, though most are Swiss in origin. *Around Switzerland in 80 Maps* is a remarkable journey through time and space, a magical history tour that uses the past to explain the present.

The book is divided into thematic sections, with the maps appearing in chronological order within each. Every map has its own text and most stand alone as a snapshot of time or place. Although many are naturally linked by events or locations, they don't always need to be read in the order they are presented. However, given that they all cover one country over a set period of time, there are some overlaps, providing different viewpoints of the same events.

The first section, **Borders & Lands**, is a collection of 12 historic maps covering almost four centuries of Swiss history. From the oldest existing map of Switzerland, published in 1480, through to the Dufour Map of 1865, which marks the birth of modern Swiss cartography, they chart the growth of the Alpine republic and its changing borders.

Town & Country looks at the regions and cities of Switzerland in more detail, with 15 maps of different scales and sizes. Two are particularly fascinating so are each spread across multiple pages: the oldest map of Lake Geneva and the Martini map of Fribourg. Switzerland may have been neutral for ages but fighting has constantly swirled in and around it. The nine maps in **War & Peace** show that the country has been both a battlefield and a safe haven during the past three centuries, and reveal a Soviet view of Basel during the Cold War.

The largest section is **Transport & Tourism**, with 21 maps tracing the development of Switzerland's famous transport system, from the first railways to the new Gotthard tunnel under the Alps. Plus a very special tourist map of Zurich that is much more revealing than most. **People & Power** looks at aspects of life in modern Switzerland that don't often appear in cartographic form. There are fourteen maps about famous Swiss products, such as cheese or watches, and the Swiss people themselves, in terms of births and deaths, votes and voices. And, of course, cows.

In **Fantasy Switzerland** eight curious maps deal with things that didn't happen, or give alternative views of the country. The medieval city that was never built or the plans for an underground high-speed train – and even a Greater Switzerland with 40 cantons.

The 80th and last map is a spectacular view of Switzerland from space, the only map in the book not created by a human hand, even if the method of capturing it was. Finally, a reference point for all the maps, especially those which distort reality by accident or by design, is the **Appendix**, which shows the 26 Swiss cantons as they are today.

Timeline

Borders
&
Lands

Four centuries of Swiss
cartography – from
the first national map
to the Dufour Map

1   The circular Island

At first it's hard to see that this oldest map of Switzerland actually is Switzerland: a stylised globe with mountains in the middle encircled by a ring of blue. But then names leap out from the dense letters, with Zug on the left and 'Berna' on the right clearly identifiable.

Although it's in the style of the circular Mappa Mundi maps of that time, this one places Mount Rigi at the centre rather than Jerusalem, as was normal then. It was first drawn in 1479 by Albrecht von Bonstetten and can be seen as a political statement, literally putting Switzerland at the heart of the world.

Back then Switzerland was not one country but a confederation of communities that had gradually united over the previous 150 years or so. Early military success against the Habsburgs had recently been reinforced by crucial victories over mighty Burgundy. The Swiss were now a force to be reckoned with and Bonstetten wanted Europe to know about "this people, their land, their customs and deeds".

He drew south ('Meridies') at the top, north ('Septentrio') at the bottom and arranged the eight cantons (known then as *Acht Orte*, or Eight Places) around Rigi: clockwise from the top 'Urania', 'Underwaldia', 'Berna', 'Lucerna', 'Thuregum', Zug, 'Clarona', 'Switia'. In today's terminology that is Uri, Unterwalden, Bern, Lucerne, Zurich, Zug, Glarus and Schwyz.

The blue ring with gold stars is remarkably similar to the European Union flag, albeit with more stars than the EU's 12. How ironic that over 500 years after Bonstetten's map, Switzerland is once again an island surrounded by a sea of star-studded blue.

Albrecht von Bonstetten (1445-1505) came from Canton Zurich and was elected dean of Einsiedeln monastery in 1469. The map first appeared in his text *Superioris Germanie Confoederationis descriptio*, published in 1480.

tes se ad angulos rectos Una earū
speculet protralji a loco solis meridi
ane et ptendi indirectū vsq; ad sep
tentrione Alia vero ab ortu solis ad
eius occasum. Hee linee diuidunt
tota Sfederatoru In quatuor ptes
p quatuor differetias positionis li
nearū vt in forma patet

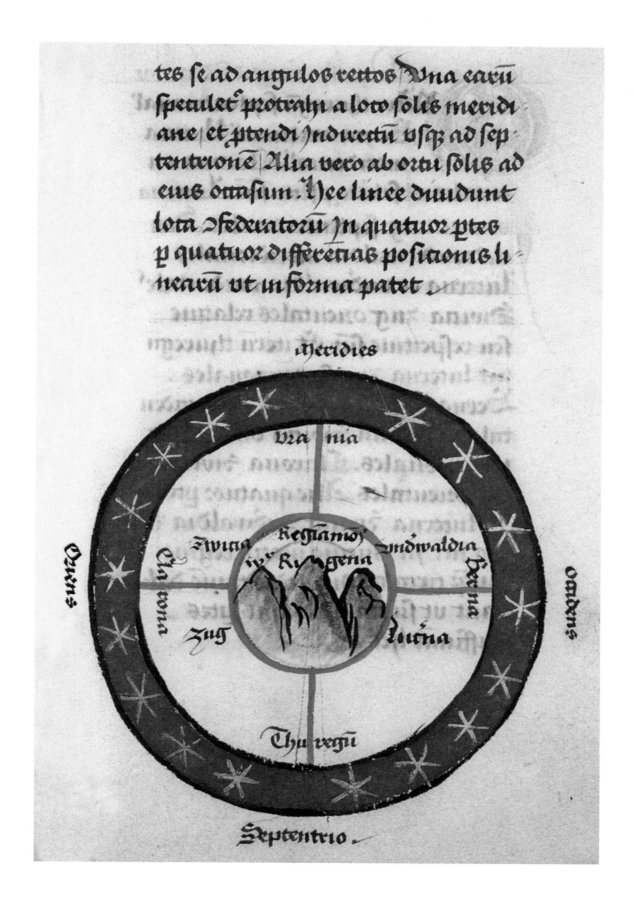

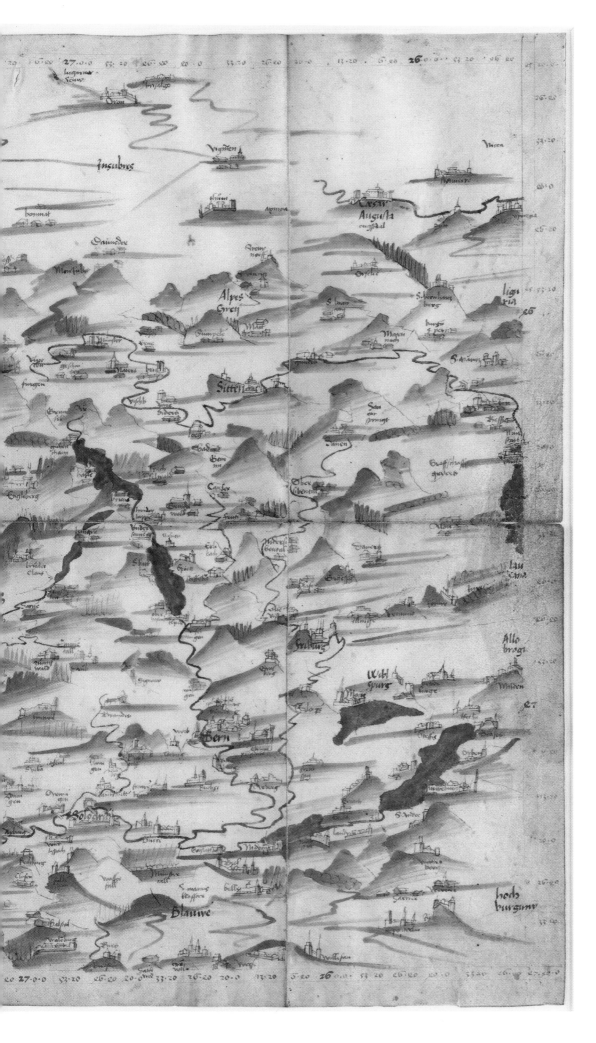

2

The first real map of Switzerland

It is the oldest surviving complete map of Switzerland, or at least Switzerland as it was then.

▶

2   The first real map of Switzerland

◄

It is the oldest surviving complete map of Switzerland, or at least Switzerland as it was then. Compared to the simplified one from 1479 (see page 14), this is miles ahead: the real first map of the country, dating from between 1495 and 1497.

From Lake Constance on the left across to Lake Neuchâtel in the bottom right, this map by Konrad Türst covers all of Switzerland at that time. That roughly relates to today's German-speaking area minus the cantons of Basel, Appenzell and Schaffhausen; although the latter two are on the map, they were not then part of the Confederation. So Ticino and western Romandie are missing, and Valais is shown only as far as Sion (written as Sitten on the map).

To our modern eyes the map is upside down. But for Türst this orientation was normal even logical, especially at a time when there were few other maps for comparison. Looking south to the Alps from Zurich, it was natural to draw a map with those mountains as the horizon. Not only that but it was the view towards Italy, the home of both Rome and the Renaissance.

Distances are not true to reality, mainly because Türst relied on travellers relating how long it had taken them to reach him. A day's walk in the valleys covers a very different distance from a day's walk in the mountains, so skewing his raw data. And, beside Lake Neuchâtel, notice how small Lake Biel is compared to Lake Murten; they should be the other way round.

It's a compelling map to pore over. You can almost hear the nib scratching on the paper as you read the place-names dotted among the green hills.

Konrad Türst (ca 1450-1503) was a doctor from Zurich who also lived in Bern. He only ever made one map, though two copies exist: a German version in Zurich, a Latin one in Vienna. It was folded into his book, *De situ Confoederatorum descriptio*.

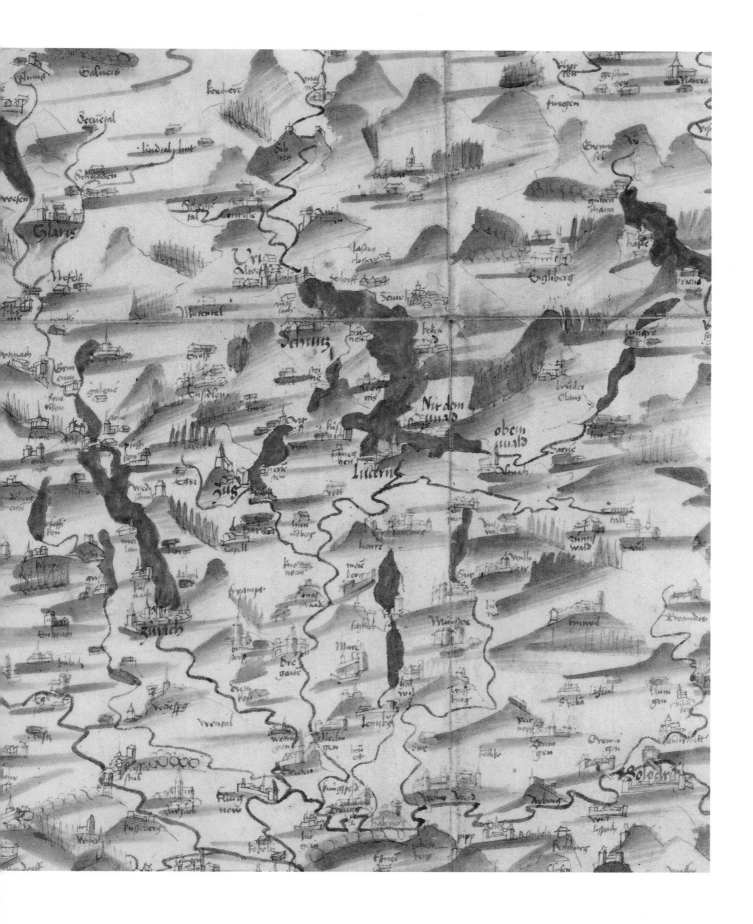

TABVLA NOVA

Hec tabula ordinata est ad sinum medium

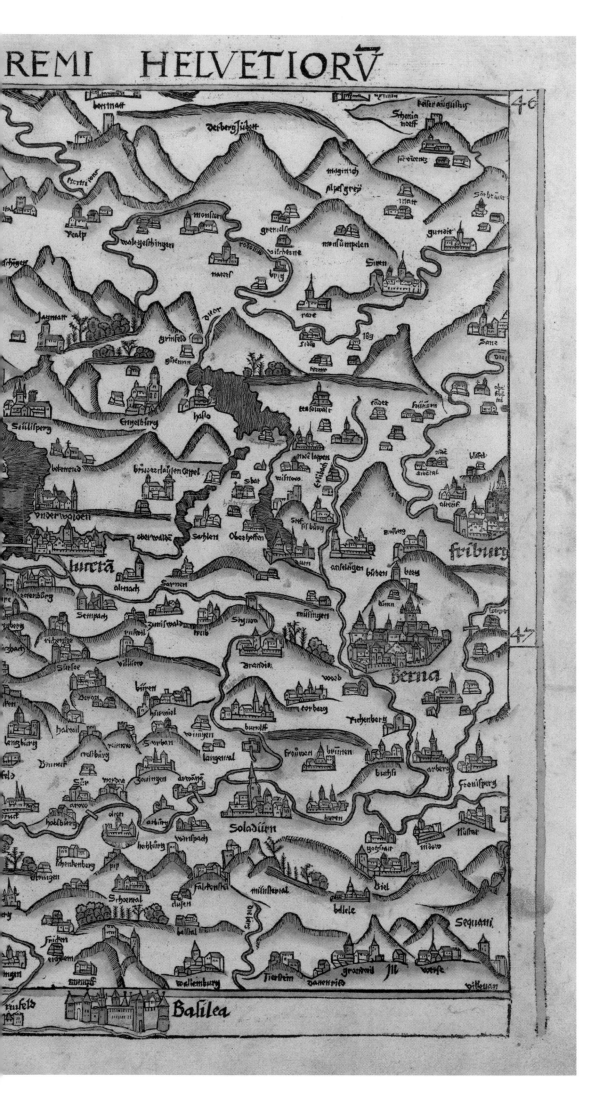

3

The thirteen cantons

The year 1513 was the start of a new era in Swiss history. With the accession of Appenzell as the 13th canton, the old Confederation was complete.

▶

3 The thirteen cantons

◄

The year 1513 was the start of a new era in Swiss history. With the accession of Appenzell (still united: see page 56) as the 13th canton, the old Confederation was complete. This would remain the situation until the Ancien Régime, as it later became known, collapsed after the French invasion of 1798. Its territory, including allies and dependencies, was almost the same then as today, with the notable exceptions of what are now Vaud, Geneva and Jura.

1513 was also the high point of Swiss military might. Victories against Burgundy in 1476-77 and the Holy Roman Empire in 1499 had secured Swiss independence, and encouraged five more cantons to join the Confederation. Ticino had been conquered and brought under Swiss control, but then defeat by the French at Marignano in Italy (1515) stopped Switzerland's expansionism in its tracks.

This map is oriented with south at the top, as was usual then, but shows no national or cantonal borders and logically doesn't include the French-speaking part as it wasn't then Swiss. Zurich is shown as 'Turegum', after its Roman name Turicum, and towns are in 3D, with the tower of Bern's Münster still clearly under construction; that part of the tower was completed in 1588.

Waldseemüller used the Türst map (see page 16) as his base but Basel had joined the Confederation in 1501, so he had to add 'Basilea' outside the bottom edge. Some names and towns along the map's sides are cut short; apparently the map was drawn larger than this and trimmed to fit into the atlas.

Martin Waldseemüller (c1472-1520) was a cartographer from southern Germany who published his atlas in Strasbourg in 1513. His famous wall map of the world in 1507 (called *Universalis Cosmographia*) was the first to use the term America.

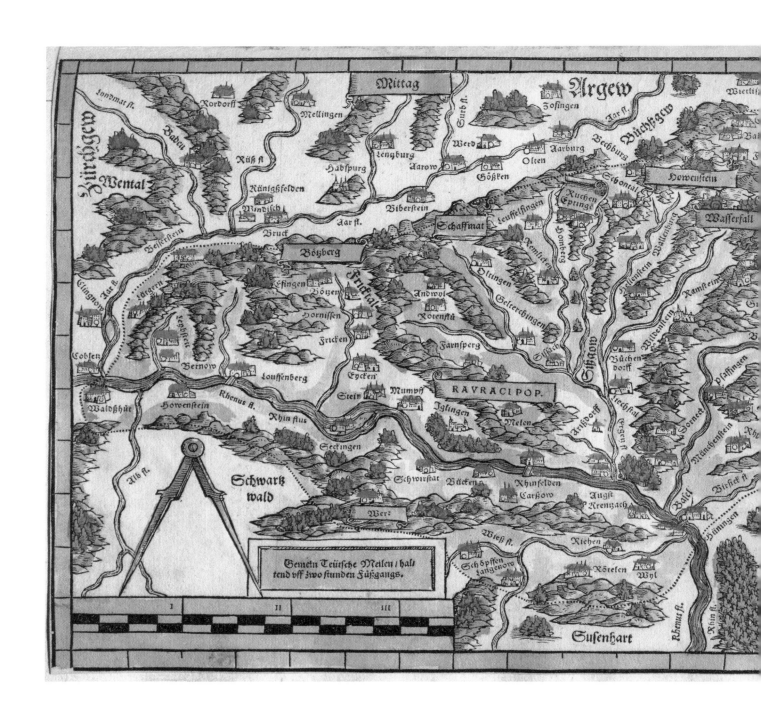

4 An atlas in all but name

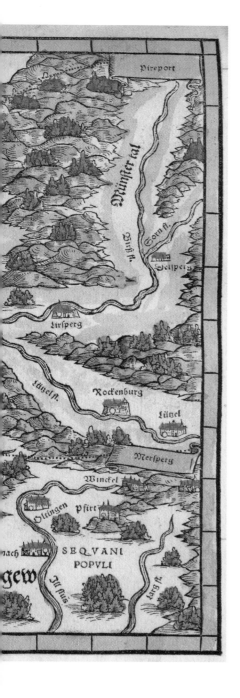

With its eight sectional maps of the Confederation, Johannes Stumpf's book of 1548 was both the first atlas of Switzerland and the first of any country in the world. Not that it was called an atlas, as that word didn't come into use until half a century later (see page 28). He called it *Landtafeln* (literally "country plates").

While there are also maps of neighbouring 'Tütschland' and 'Franckrych', it is the detailed Swiss ones that are the most fascinating, as shown by this map of the *Rauracer/Basler Gelegenheit*. Its title refers to both Basel and the Rauraci, an ancient tribe conquered by the Romans but whose name still lives on in the ruins at Augusta Raurica.

Many names have stayed the same, such as Fricktal and Baden on the left-hand side (which is east as this map is south oriented); others have subtly changed so that they may sound the same but are spelt differently, either simply swapping double consonants – 'Schaffmat' is now Schafmatt – or rewriting vowel sounds, eg 'Howenstein' to Hauenstein. Or even both changes, so that it is 'Louffen' not Laufen.

Then there are the names that have disappeared altogether. Modern maps have no 'Buchsgew', 'Sissgew' or 'Zürchgew', though 'Argew' is now Aargau. The term gew (also spelt gow, see page 54) meant an administrative district and has survived as gau, but not just in Switzerland. 'Suntgew' in the bottom right is now Sundgau, a region of Alsace.

Stumpf created the maps from woodcuts. All were printed in black and white, but some were coloured in afterwards, and then published in several editions, both loose and bound.

Johannes Stumpf (1500-1577/78) was a German priest who moved to Switzerland, preached the Reformation alongside Zwingli, and married three times. The *Landtafeln* were originally maps taken from Stumpf's landmark history of the Confederation, published in 1547-48.

5   The national borders appear

As of 1570 Switzerland as an entity was finally
on the map, thanks to Aegidius Tschudi. His Helvetia
descriptio might look upside down but it displayed
the country of Switzerland for the first time, with
national borders and coloured shading to distinguish it
from its neighbours.

This map stretches from 'Basilea' and 'Schafhusen'
in the north (ie bottom) to 'Lacus Verbanus' (now Lake
Maggiore) in the south, plus most of the French-
speaking parts – the Bernese had conquered Vaud in
1536. Geneva and the western end of 'Lemanus Lacus'
are beyond the Swiss border, which can clearly be
seen as the yellow-pink line near 'Losanna'.

The Swiss lakes don't all look quite as they should.
Lake Lucerne is two separate entities, while Lake Biel
almost disappears completely: it's the shrivelled appen-
dage to Lake Neuchâtel (or 'Neoburgensis Lacus') in the
bottom right part of the map. But the map is easier to
read than its predecessors, including Tschudi's earlier
Swiss map of 1538, which wasn't in colour and showed
no borders.

This version was produced by Abraham Ortelius
in Antwerp. A friend and rival to the famous Mercator
(see overleaf), he created an early world atlas,
although that was Mercator's term not his; he called
his *Theatrum Orbis Terrarum* (or Theatre of the
Whole World).

Whereas Mercator drew his own maps, Ortelius
used ones already available and re-engraved them for
use in his book. He copied Tschudi's map, shrinking
it to fit into his *Theatrum* and keeping the southerly
orientation, despite other maps in his collection having
north at the top.

Aegidius Tschudi (1505-1572) was a soldier and politician from a famous Glarus family.
He is best known as a historian who wrote the first drafts of *Chronicon Helveticum*
(eventually published 1734-36), a book which helped popularise the William Tell legend.

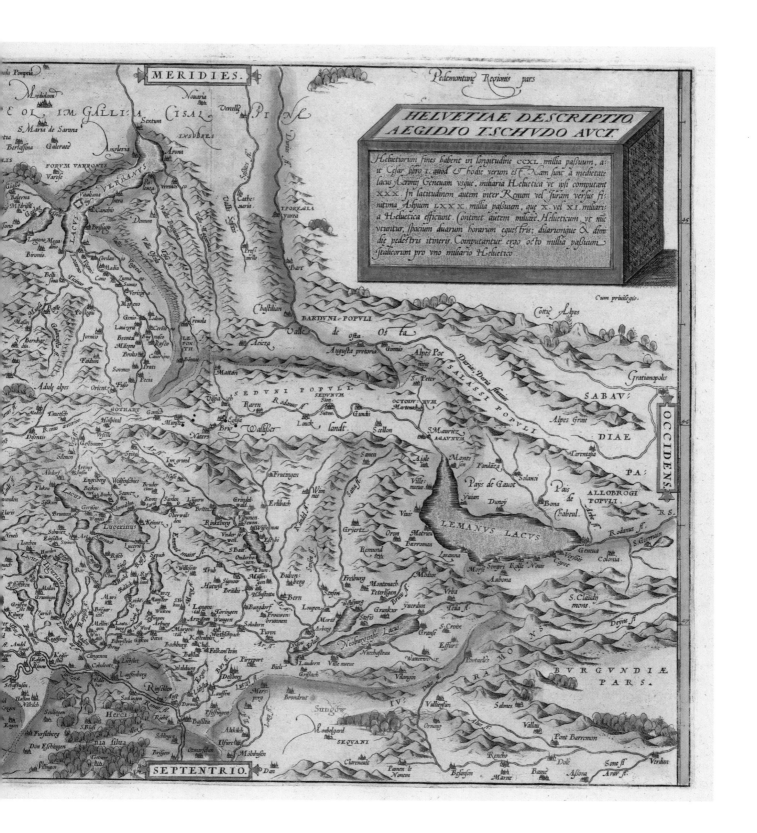

6 Switzerland almost as we know it

North at the top, borders clearly shown and topography close to reality: this Mercator map is one of the first to show Switzerland – or 'Helvetia' as he calls it – as we know it today. Almost.

For the first time there was a map that we can look at now, over 400 years later, and easily identify without us squinting or turning it round. As with Tschudi (see previous page), Vaud and Valais are on the map but this one has both the orientation and proportions we are used to. It is the area now known as Switzerland in a state we can recognise.

Gerardus Mercator wanted consistency and scientific principles to govern his maps. After the success of his famous world map in 1569, he turned to mapping Europe in more detail. His bound set of 51 maps was first published in 1585 and all were oriented with north at the top.

The collection was updated and expanded for a new edition ten years later, published by his son after Mercator's death at the age of 82. That 1595 'atlas', a term Mercator coined, contained 107 maps including this one of Switzerland, all engraved by hand and, for the more expensive version, coloured afterwards.

Mercator travelled widely but never came to Switzerland. He relied on others' maps and accounts as his sources, which are all listed on the back, to create maps that were scientifically researched. This Swiss map is remarkably accurate but it makes the same mistake with Lake Lucerne as Tschudi did – dividing it in two – though Mercator at least got Lake Biel right.

Gerardus Mercator (1512-94) was born in Flanders and went on to become the most famous name in cartography. His 1569 projection of the world - still used today - was revolutionary for making navigation at sea much easier. This Swiss map is the 2nd edition from the 1595 atlas.

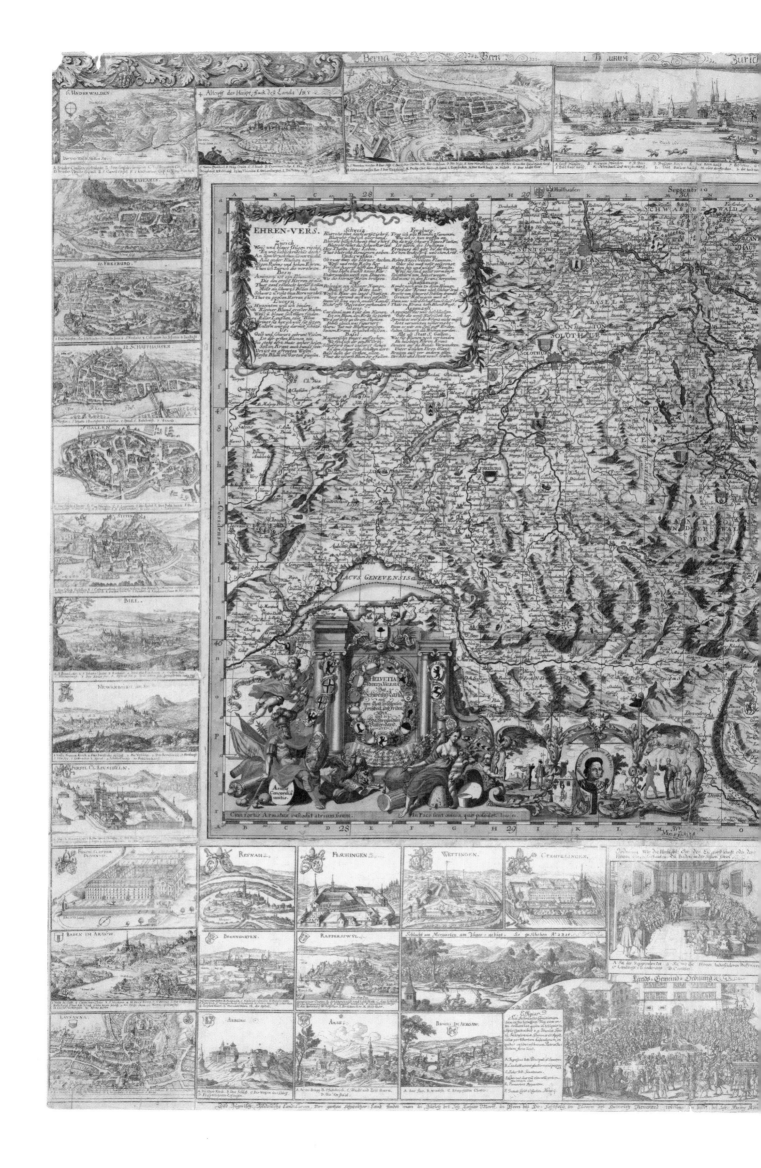

7

A unique map of maps

Heinrich Muoss only ever made one map but he made sure it was a memorable one. And a very big one: the whole thing measures 101cm by 87cm.

▶

7   A unique map of maps

◀

Heinrich Muoss only ever made one map but he
made sure it was a memorable one. And a very big one:
the whole thing measures 101 cm by 87 cm.

At the centre is the colourised map of Switzerland
that clearly shows the cantons' borders and coats of
arms, as well as details like Zurich's fortifications
and lovely vignettes, such as the portrayal of William
Tell shooting the apple off his son's head. Add in the
52 places shown in detail around the edge (see overleaf)
and it is a unique view of Switzerland at the beginning
of the 18th century.

Most interesting is the map's title - *Helvetia, Rhaetia,
Valesia: das Schweitzerland* – and not just for the spel-
ling of Switzerland. The Confederation is shown as
three separate but allied parts: Helvetia (ie the thirteen
cantons), Rhaetia (the old name for Graubünden) and
Valesia (ie Valais). In 1698 Switzerland as a single entity
was still not a reality but it had secured its complete
independence.

The Treaty of Westphalia (1648) had ended the Thirty
Years' War and recognised Switzerland's full independence from the Holy Roman Empire, but that didn't
mean its borders were the same as today, either inter-
nally or internationally. Switzerland was made up of the
cantons plus various dependent territories, such as
Vaud or Aargau, and independent allies, eg Geneva,
shown here as distinct but separate from the rest of the
country.

▶

urcuty. D. Der Spital. E. Das Raht-hauße. F. Das

1. Munſter. 2. S. Iohañs 3 Barfüſſeren. 4. Cloſter. 5. Spital. 6. Rahthauß. 7. Vnnoſt.

oß Valeria. Das Schloß der Maiorey. vnd Bischoffliche Reſi
F. S. Theodori. G. Rhodanus fluius H. Der Spital.

The 52 individual views, drawn separately and pasted on afterwards, make the Muoss map uniquely fascinating. Alongside the likes of Schaffhausen, 'Lausanna' and Bern are smaller places such as St Urban or 'Arburg', meaning that for many locations shown here this is the earliest surviving portrayal.

Among the towns are also monasteries, such as the one at 'Dissentis', which seems rather odd at first but isn't when you realise that they were also allies. Each *'Fürstliche Closter'* (or princely monastery) was still independent of Swiss control, often ruled by a powerful Prince-Bishop who was more regal than religious.

A closer look at the elaborate title crest reveals the ring of thirteen cantonal coats of arms in the centre plus the shields of allies, such as Neuchâtel, Biel, Rottweil and Mulhouse. Many of those places also appear among the detailed black-and-white town plans around the edge of the map. They are what make this map a masterpiece of history.

◀

Heinrich Ludwig Muoss (1657-1721) was an artist, printer and bookseller from Zug. His map of 1698 was updated and coloured for a second edition in 1710 (shown here). A third edition was issued in 1770, long after his death.

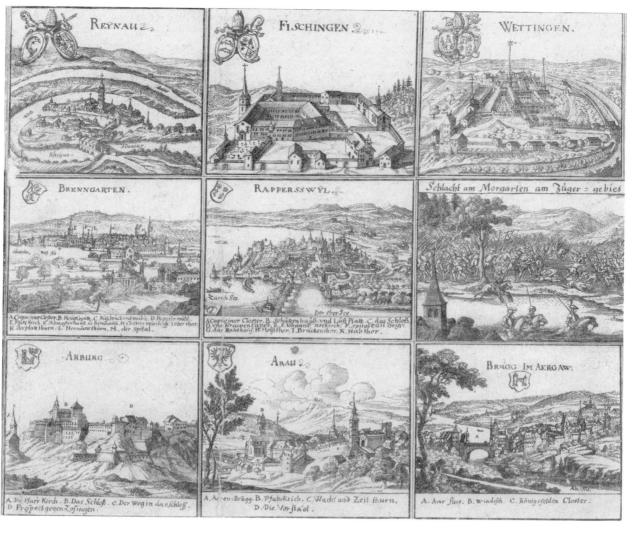

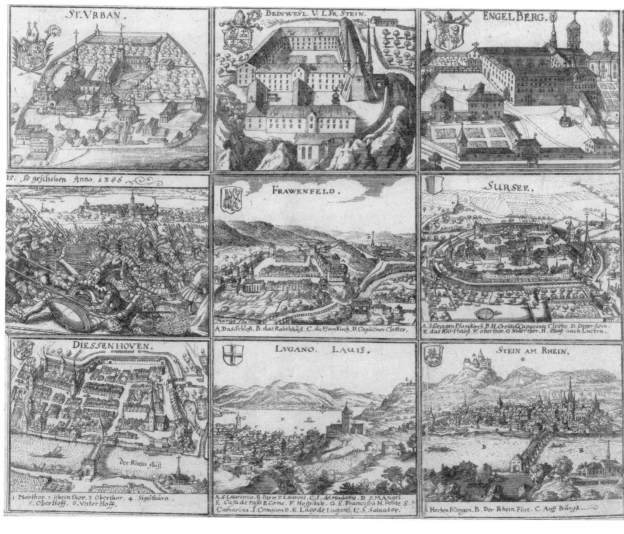

HELVETIAE PAGI seu CANTONES. XIII. OHRT der EIDGENOSSSCHAFT. 1. ZURICH. 2. BERN. 3. LUCERN. 4. URI. 5. SCHWEIZ. 6. UNDERWALDEN. 7. ZU

NOVA
HELVETIAE
TABULA GEOGRAPHICA

8

A landmark of scientific cartography

At 41,285km² (present-day area) Switzerland isn't a large country, and never has been. It could easily fit onto one map but its landscape is so convoluted that showing any detail can be hard.

▶

8   A landmark of scientific cartography

◀

At 41,285km^2 (present-day area) Switzerland isn't a large country, and never has been. It could easily fit onto one map but its landscape is so convoluted that showing any detail can be hard. The answer? Divide the country into quarters and create four maps, each the size of one normal one.

That's exactly what Scheuchzer did in 1712. His four sheets each measured 73cm by 54cm, and were separately drawn and engraved. When assembled, the whole map was roughly 1.5 metres by one metre. In other words, huge.

The black-and-white original was so accurate and became so famous that it was copied many times. This version, from Schenck in Amsterdam in 1715, is one of the clearest, with the cantons and different constituent parts of Switzerland plainly shown. This section is of south-eastern Switzerland covering what would become Graubünden and Ticino (still divided into separate districts).

Around the map's edge are scenes by artist Johann Melchior Füssli, showing Swiss life, industry and landscape, including the Rhine Falls. Across the top are the names of the thirteen cantons plus the assorted allies, dependencies, and protectorates that constituted Switzerland at that time. It was a complicated state of affairs that would last until 1798 (see overleaf).

Together with the Muoss map of 1710 (see page 32), Scheuchzer provides a fascinating picture of Switzerland at the start of the 18th century. He also set the standard for a scientific approach to cartography, with his very precise map becoming a yardstick for ones that followed.

Johann Jakob Scheuchzer (1672-1733) was a Zurich man with many talents: physicist & physician, historian & geographer, writer & publisher, polymath & palaeontologist, mountaineer & mapmaker. He wrote prolifically about the Alpine landscape and nature.

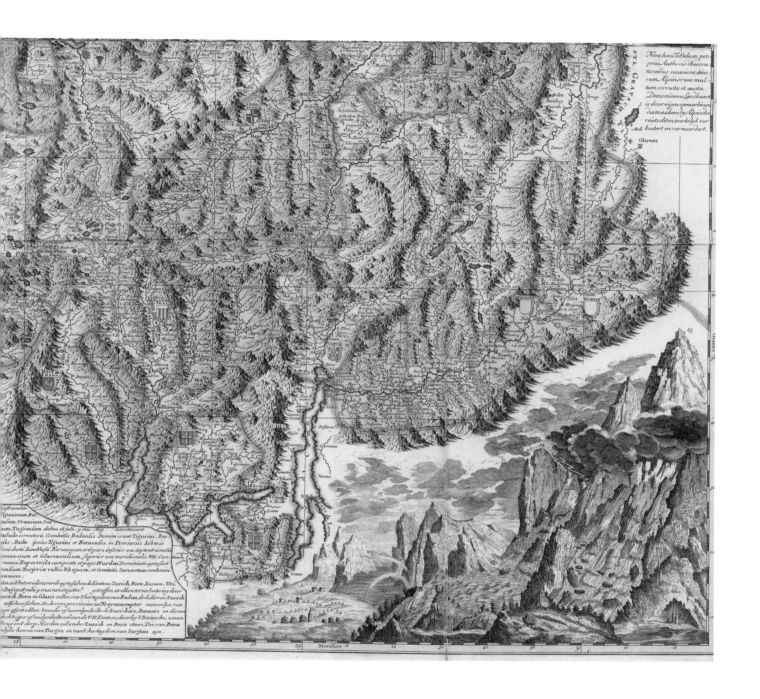

9 The Helvetic Republic

In 1798 Switzerland was conquered for the first and only time in its history. The French revolution, then invasion, precipitated the complete collapse of the Ancien Régime of thirteen cantons with their satellites, dependencies and allies.

The French occupiers first annexed Geneva and the bishopric of Basel (now Canton Jura), then imposed a central unitary state on what was left. The Helvetic Republic was declared on 12 April 1798 and its flag (a red, green and yellow tricolour) soon flew over its first capital, Aarau.

A new constitution reformed the whole political system and reorganised the internal structure. Gone was the historic but messy configuration that had grown since Switzerland's foundation. In its place came 18 shiny new cantons, some with new names, most with new borders.

Six old cantons (Basel, Fribourg, Lucerne, Schaffhausen, Solothurn and Zurich) were joined by new ones from former dependencies: Vaud – renamed Léman – and Thurgau plus Aargau and Ticino. The latter pair were both split in two cantons (Aargau and Baden, and Bellinzona and Lugano respectively), as was Bern (into Bern and Oberland). The new Republic of Valais was included until its independence in 1802. Totally new cantons were Waldstätten (shown in blue), Linth (in orange), and Säntis (in green).

This map shows the 18 cantons in 1799, just before Rhaetia (present-day Graubünden) joined in April of that year but without its dependent territory of Valtellina, which it had been forced to cede to the new Cisalpine Republic. Four years later the Helvetic Republic was abolished.

The map appeared in the first edition of *Helvetischer Almanach*, a political periodical that detailed the state of the nation. It was published by Orell Füssli from 1799 to 1822, and in later editions each canton had its own map (see page 82 for Ticino).

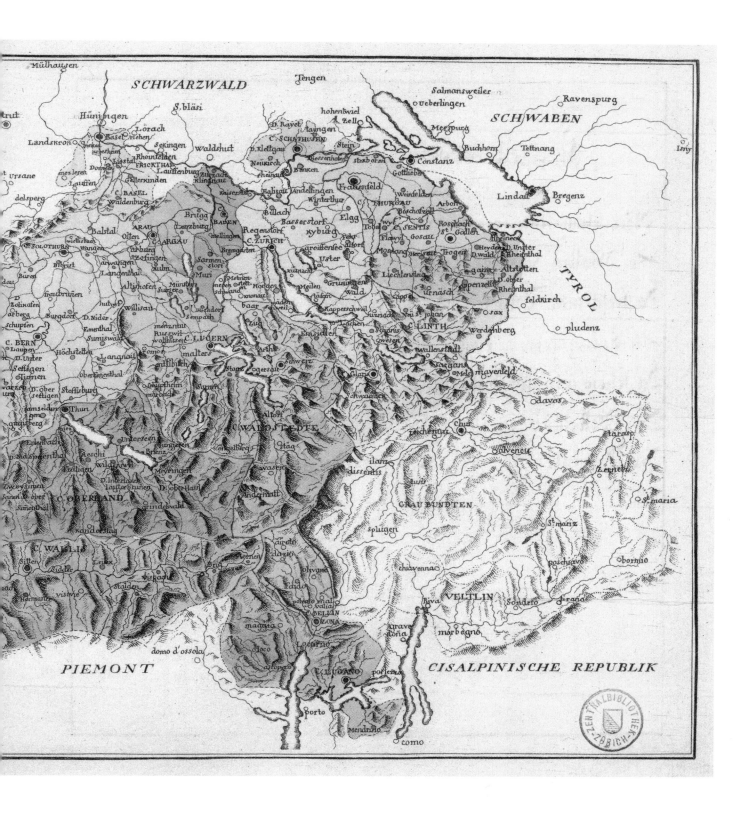

10   A temporary mediation

The ink had barely dried on maps of the Helvetic Republic (see previous page) when it was abolished. The French-imposed experiment in central government had not worked and the country slowly descended into chaos. Worst of all, Switzerland was no longer neutral but a satellite of France: it became a battlefield for the great European powers (see page 92) and thousands of its men were dragooned into the French army.

In 1802 Napoleon, now in charge in Paris, withdrew French troops from Switzerland but the ensuing disorder forced him to intervene again quickly. The result: another new Swiss state. The Act of Mediation came into effect on 19 February 1803, replacing the unloved and unmissed Helvetic Republic.

The new Switzerland was still under French influence but was now a revamped confederation, a compromise between centralism and federalism. Alongside a single central assembly (but no capital city) were the 13 original cantons from the old Confederation (see page 20), plus six new ones: Aargau, Graubünden, St. Gallen, Thurgau, Ticino and Vaud. Geneva and Jura remained French, and Valais was nominally independent, at least until it was also annexed by France in 1810.

All this change was a cartographical nightmare, hence the text on this British map explaining Switzerland as "now comprised in 19 cantons, beside the Allied republic of Valais, the County of Neuchâtel Subject to Prussia, the Valteline annexed to Italy, and the Principality of Porentru incorporated with France."

Mediation lasted as long as Napoleon and died in the wreckage of his failures.

This English map has precise details of its publication printed on it, as was normal in Britain at that time: "Published July 1 1808 by Robert Wilkinson, No 58 Cornhill, London". The explanatory text was as convoluted as the Swiss political situation.

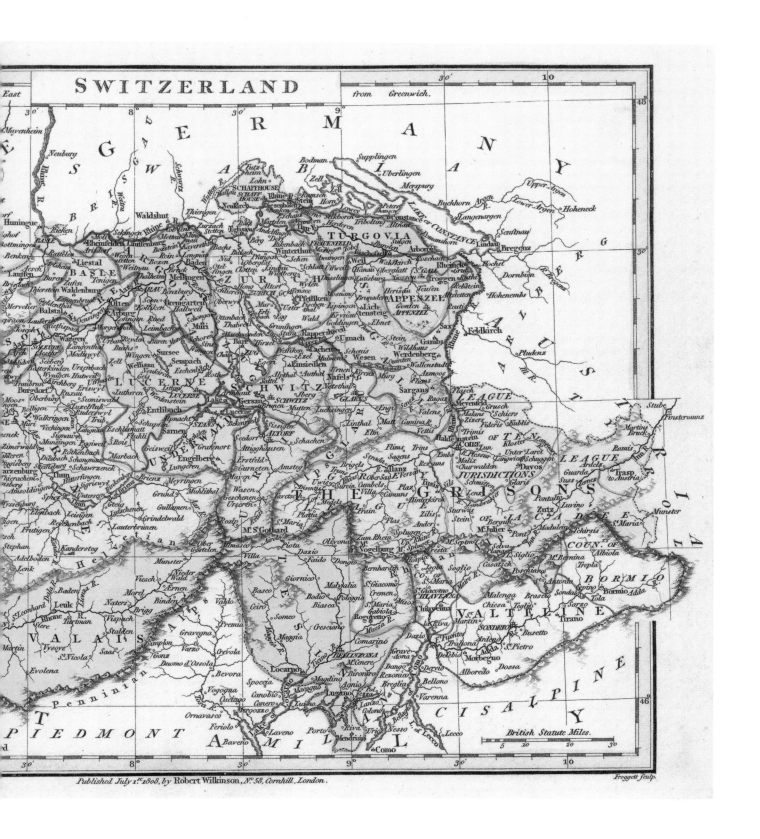

SWITZERLAND

from Greenwich.

East

Published July 1st 1808, by Robert Wilkinson, No. 58, Cornhill, London.

Froggett sculp.

11   1815: a neutral modern country

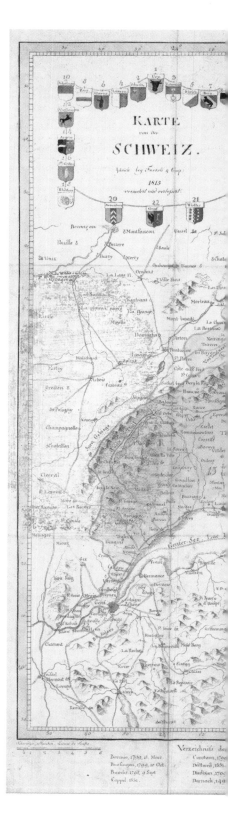

With the defeat of Napoleon, Switzerland was reinvented again, its fourth incarnation in under 20 years. But this latest form was one that would last until today, at least in terms of Swiss national borders and neutrality. Both had been violated during the Napoleonic Wars but both would be secured in 1815.

At the Congress of Vienna, the territory annexed by France was restored in the form of three new cantons – Geneva, Neuchâtel and Valais – though Geneva had no land border with Switzerland, so acquired some French land to create a bridge. The former bishopric of Basel (now Canton Jura) was given to Canton Bern in compensation for losing Aargau and Vaud. Valtellina was lost forever.

The Congress declared on 20 March 1815 that Swiss neutrality should be recognised and guaranteed by all powers. On that same date Napoleon left Elba to try and conquer the world in 100 days; Switzerland responded by invading Burgundy, its last ever aggressive foray abroad. It withdrew again shortly afterwards. Swiss neutrality was officially confirmed by the Great Powers on 20 November.

Switzerland's internal political structure was reorganised yet again. On 7 August the Federal Pact created an alliance of 22 sovereign cantons. Known as the Swiss Confederation, it would last until federation in 1848, although the name continues on today.

Since then some cantonal borders have changed – Basel's division in 1833 and Jura's independence from Bern in 1979 – but Switzerland's national borders have remained as constant as its neutrality. Both survived the 20th century intact, unusually for a European country.

This 1815 map from Fuessli & Comp. in Zurich shows all 22 cantons, numbered in order of joining the Confederation. Major historic battles are listed across the bottom: the "recent" ones have the exact date as they were within living memory.

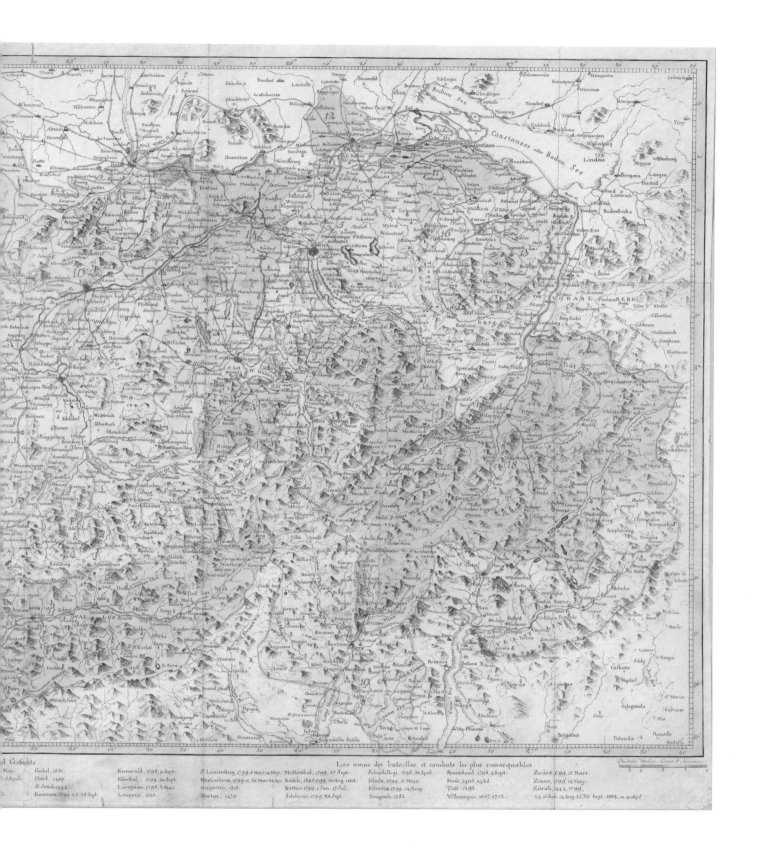

12   The godfather of modern Swiss maps

In 1832 the Federal Topographic Bureau started the mammoth task of accurately measuring every centimetre of Switzerland. In charge was Guillaume-Henri Dufour. The result was the first official series of Swiss maps, collectively known as the Dufour Map.

It is not one map but a series of 25 separate sheets, each one 70cm by 48cm and to a scale of 1:100 000. The first edition took twenty years to complete, with the final sheet published in 1865. This sheet, number 23, covers part of Valais and the border with Italy. In the upper example (first edition, 1862) Switzerland's highest mountain was still prosaically called '*Höchste Spitze*' (or Highest Point). Its 4634-metre summit was first conquered in 1855 by a British-Swiss team.

In January 1863, two months after the first edition, the Swiss government renamed the peak Dufourspitze (Pointe Dufour in French or Punta Dufour in Italian). Only 214 copies of Sheet 23 have *Höchste Spitze* on them. The lower example, dating from 1926, has the new name and also blue colouring for the glaciers; that second colour was added in 1908. Red, green and brown appeared on later editions. In 1965 the last edition of the last sheet of the Dufour Map was published.

The Dufour Map shows place names in their own language, so it is Genève not Genf, Solothurn not Soleure. This practical solution to Switzerland being multilingual is one used by most Swiss maps ever since.

In 2014 the Ostspitze, a neighbouring peak to Dufourspitze only two metres lower in height, was officially renamed Dunantspitze, in honour of the founder of the Red Cross, Henri Dunant.

Guillaume-Henri Dufour (1787-1875) grew up in Geneva and was a man with many hats: engineer, general, politician and co-founder of the Red Cross (with Dunant).
He commanded the victorious Federal army during the Sonderbund War of 1847 (see page 96).

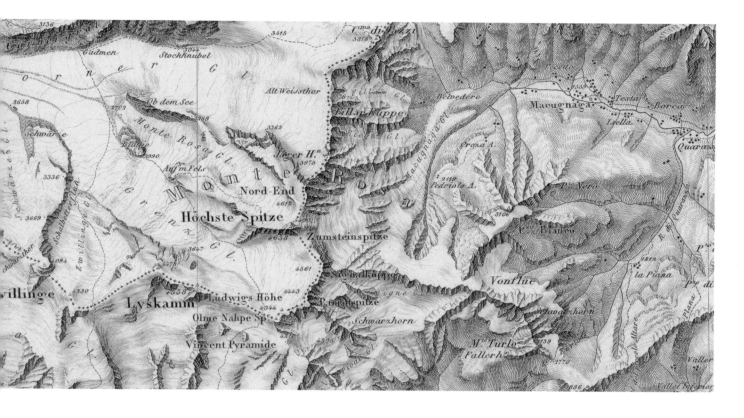

Town
&
Country

Historic maps of the
cities and regions

13

The oldest map of Lake Geneva

It's hard to believe that this amazing map of Lake Geneva was almost lost forever.

▶

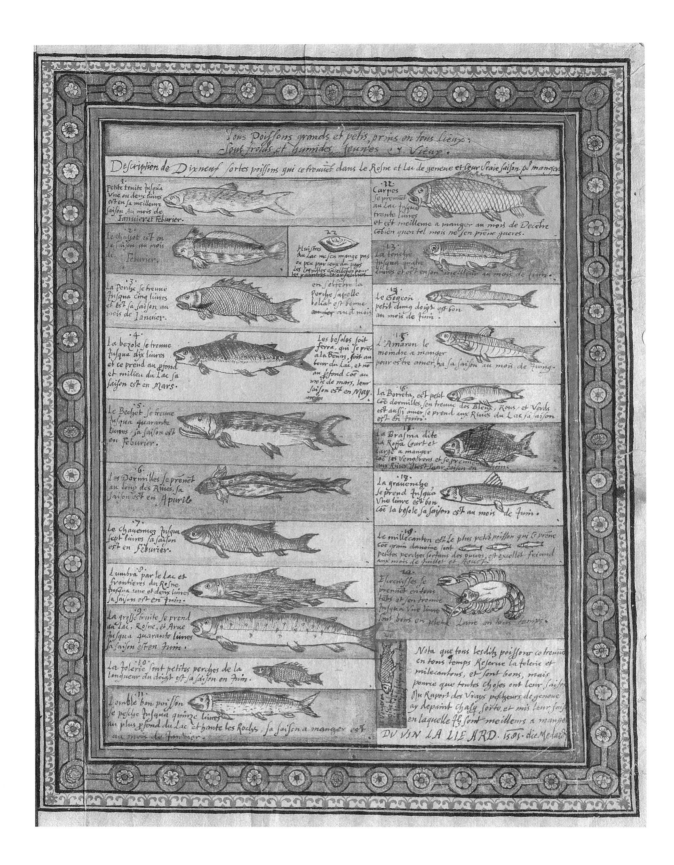

13   The oldest map of Lake Geneva

◄

It's hard to believe that this amazing map of Lake Geneva was almost lost forever. For years it had sat hidden – and forgotten – at the back of a book in the Geneva city library, and was only discovered by accident in 1827, almost 250 years after it was created. This oldest surviving map of the lake region is now carefully stored in the same library.

Drawn by a local man, Jean Duvillard, the map has a southern orientation, which makes the lake look like an upside-down croissant. The intricate detail includes clearly illustrating towns such as Evian or Nyon with red-roofed turrets, though strangely neither Geneva nor Lausanne is featured. There are even fighting chamois up on the slopes of the Dents du Midi, the blue mountains near the top of the map.

At that time Lake Geneva formed the south-western border of Switzerland. After defeating Savoy in 1536, Canton Bern ruled over Vaud and, along with Valais, also briefly controlled the southern lakeshore before having to hand it back to Savoy. The independent republic of Geneva was a Swiss ally.

Attached to the end of the map is an equally remarkable section with Duvillard's illustrations of 19 types of fish that were then to be found in the Rhone (spelt the old way, 'Rosne') and Lake Geneva. He also gives details on the best season for eating each of them, be they carp, perch or trout.

Measuring 102cm by 33cm, including the fish, the whole document is perfectly shaped to display the lake in its entirety. The actual map dates from 1588 whereas the piscatorial drawings are seven years older.

Jean Duvillard (1539-1610) was from a wealthy Geneva family, and joined the army to serve the Holy Roman Emperor Ferdinand I. In 1587 he was elected to be a *syndic*, a position at the highest level of government in medieval Geneva.

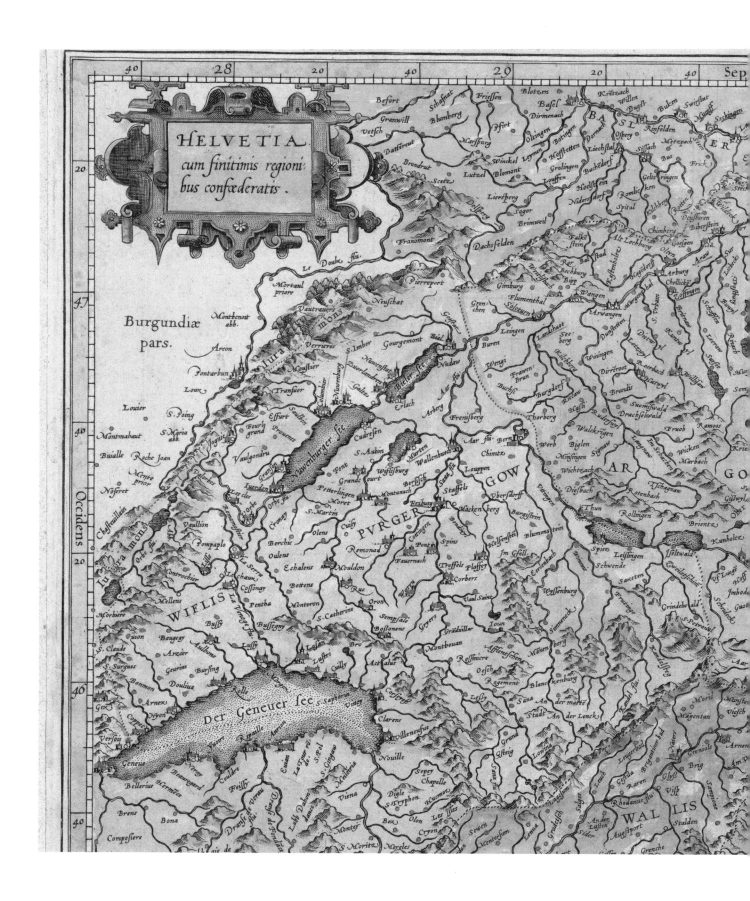

14   The lost region

Wiflispurgergow (often also spelt Wiflisburgergow) is a name that has disappeared from the map of Switzerland, but it was once on almost every one. So where was this land that maps forgot?

On this 1595 map from Mercator you can clearly see it shaded in pink and covering most of western Switzerland. The dotted-line border runs from France past Biel and Bern to Lake Thun and beyond Grindelwald, then down into Valais and round through Lake Geneva. It encompasses all of French-speaking Switzerland - the area now called Romandie - and half of the Bernese Oberland.

Its odd name comes from Wiflispurg, the old German name for the town of Avenches in Canton Vaud. On this map it is spelt 'Wiflisburg' and is just to the right of 'Nuwenburger See' or Lake Neuchâtel. The word *gow* is a medieval term for an administrative region or district that was larger than a canton but not quite a country.

It wasn't the only *gow* in town. The neighbouring province of Argow included much of the cantons of Bern and Lucerne, while further east were Zürichgow and Turgow. In its modern form of *gau*, the word can still be seen in the Swiss cantons of Aargau and Thurgau, plus Freiburg im Breisgau (in Germany).

This regional section shown here was part of Mercator's general Swiss map (see page 28), one of the first with north at the top and distances in proportion. Mercator never came to Switzerland but used his network of sources for information, which may account for oddities like 'Losan' on the north shore of 'Der Geneuer See'.

Gerardus Mercator (1512-94) is perhaps the most famous name in cartography. His 1569 projection of the world is still used today and he was the first to use the term 'atlas' for a bound collection of maps. This map is the 2nd edition from his 1595 atlas.

15   A peaceful religious division

Appenzell is not a big canton, with a total area of 415.4km², so it's perhaps surprising that it is in fact divided into two unequal halves, Innerrhoden and Ausserrhoden. The former has barely enough people to fill a football stadium, making it the least populous canton, so why are there two Appenzells?

The answer is the Reformation, which divided Switzerland as much as anywhere in Europe. Starting in 1519 with Ulrich Zwingli in Zurich, the Protestant wave swept through Bern and Basel, while most of central Switzerland remained faithful to Rome.

This war of words became a battle for hearts and souls – and lives when armed conflict broke out in 1531. The Catholics won a famous victory at Kappel, killing Zwingli, and an uneasy peace followed with Protestantism arriving in Geneva but the Counter-Reformation restoring the balance in central Switzerland.

Appenzell also split between the two faiths, but peacefully. In 1525 the canton had decided that each parish could vote on the matter, with Ausserrhoden becoming mainly Protestant, Innerrhoden staying Catholic. The compromise lasted until 1597 when the two sides voted to separate and the half-cantons were born. Under the federal constitution of 1999, half-cantons are officially equal to whole cantons, apart from in a referendum and for seats in the smaller house of the federal parliament.

That Appenzell remains split is not down to religion as that's no longer a factor in Swiss politics; the biggest issue today is that Catholic cantons get more public holidays than Protestant ones. The two Appenzells remain divided because that's how they want it to be.

Gabriel Walser (1695-1776) was both Appenzeller and Protestant cleric. His map from 1740 shows Ausserrhoden (written as 'Aussere Rooden') in red and Innerrhoden in green, with south at the top. The key details which churches are Reformed and which Catholic.

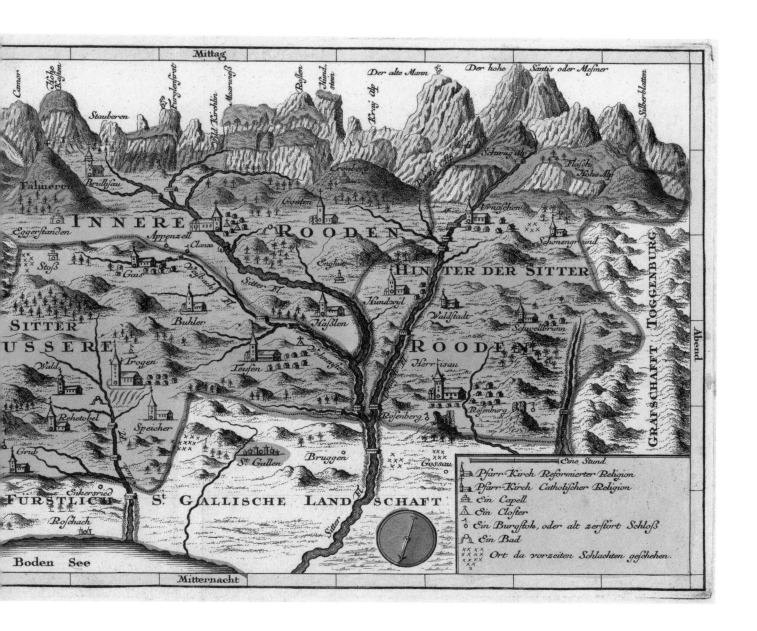

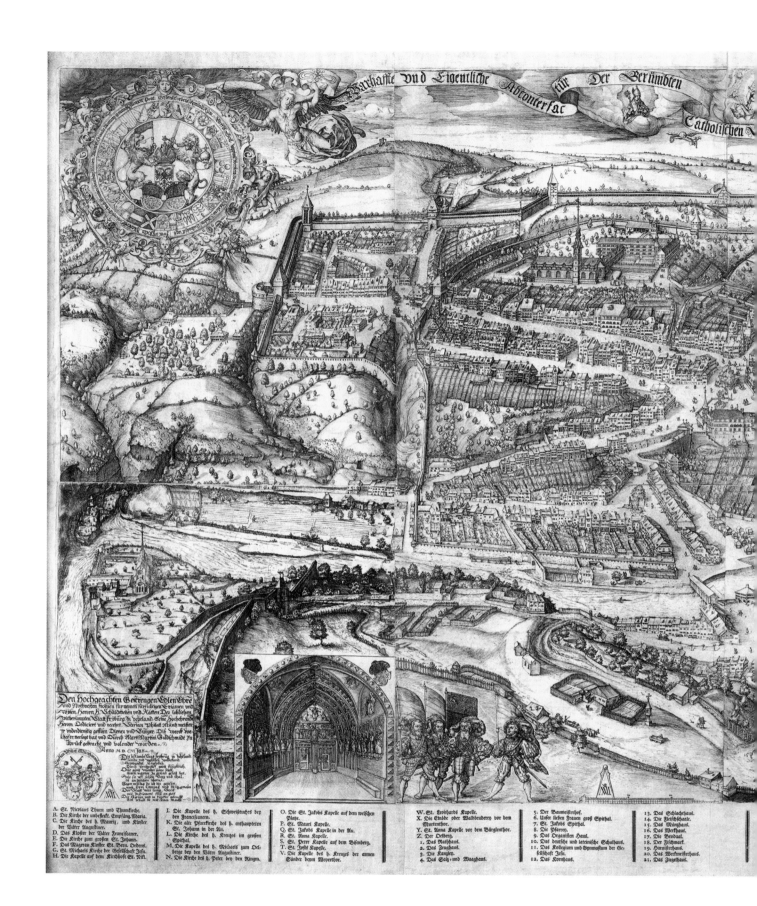

16

Magical medieval Fribourg

Welcome to Fribourg in 1606!

▶

16   Magical medieval Fribourg

◀

It's a bird's-eye view but that bird must have been an eagle because the detail is impressive. Not only the houses and gardens but the people too, from the well-dressed ladies walking a dog beside the cathedral to the gentlemen with capes and swords. Welcome to Fribourg in 1606!

This large map (156cm by 86cm) was engraved into eight copper plates by Martin Martini and shows the city as seen from the south, stretching out along the cliff above the river. His original plates, and map, now reside in the cantonal museum in Fribourg.

The population of Fribourg was then around 5,000, and it remained steady until the mid-19th century, when the city expanded. Its walls were partially demolished, the train between Bern and Lausanne arrived in 1862 and the university opened in 1889. But the old town hasn't changed much since Martini's day, although his map isn't accurate down to the last millimetre. For example the tower of St-Nicolas' Cathedral appears taller than in reality (76 metres), as are the city walls and towers.

Fribourg was founded by the dukes of Zähringen in 1157, though it soon passed to the Kyburg dynasty. In 1277 the city was sold to the Habsburgs, then was part of Savoy, before finally becoming a Swiss canton on 22 December 1481 – the ninth to join the Confederation but the first bilingual one.

German was originally the official language and it stayed that way until the French army arrived in 1798. Today the city is officially bilingual, with 63% of the population speaking French as a first language and 29% speaking German; the rest is made up of many different languages.

Martin Martini (c1565-c1610) was from Graubünden and is famous for two city maps: Lucerne (1597) and Fribourg. For the latter he received 30 livres (about 1500 Swiss francs today) and citizenship of Fribourg; enough that he drew a new map in 1608.

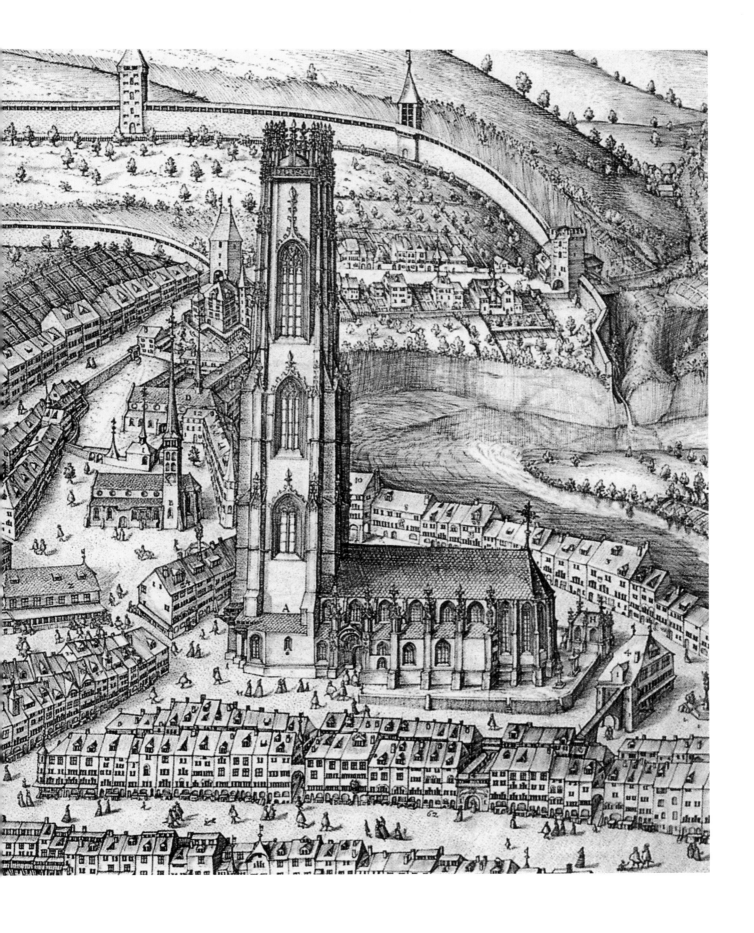

17   The republic of Valais

Switzerland's third largest canton has been part of the Confederation since 1815 but the road to accession wasn't easy. Valais was a republic, then a canton, then a republic again, then a French département, before finally becoming the 20th Swiss canton.

This beautiful map from 1682 shows Valais as an independent republic separate from but allied to 'Schweitzerlandt', as written across the top. Valais (Wallis in German) was historically organised into districts, or *Zenden* in German and *Dizains* in French. The seven shields on the map represent the original districts, from the cantonal capital of Sitten (Sion in French) to Goms in the east.

The canton has always been divided between the French-speaking Lower Valais and the German-speaking Upper Valais to the east. For many centuries Upper Valais had the upper hand, and it was on the pretext of liberating the French-speakers of Lower Valais that France invaded in 1798. Valais was incorporated into the new Helvetic Republic (see page 40), until it seceded four years later to become a nominally independent republic.

In 1810 Valais was annexed by France as the département of Simplon but, following Napoleon's defeat, it re-joined Switzerland on 4 August 1815. The number of *Zenden* rose to 13, hence the modern cantonal flag having 13 stars, as opposed to the seven shown on these shields in the bottom right.

Today Valais is one of three officially bilingual cantons, with French speakers in the majority (67%). Many towns have two names, so Siders, Visp and Leuk in German are Sierre, Viège and Loèche in French.

This map was drawn in 1682 by **Antoine Lambien** but only published in Lyon in 1709. It shows the seven historic districts – Sitten, Siders, Leuk, Raron, Visp, Brig, Goms – which have since been joined by Monthey, St-Maurice, Entremont, Martigny, Hérens, & Conthey.

18   The lion of Zurich

Think of an animal associated with Switzerland and you'll most likely picture a cow or a goat, maybe a bear or an ibex. Perhaps a marmot. But probably not a lion. It's not as if there have ever been any in Switzerland, except in a zoo; even Hannibal only brought elephants with him.

But the lion is a popular emblem in heraldry, not just because he looks good on a shield but because he is the king of the beasts. No surprise then that he became the emblem for the city and canton of Zurich, which is the largest member of the Swiss pride. The *Zürileu* (or Zurich lion) first appeared on the official coat of arms in 1490 and has remained ever since.

He also pops up as a statue all over the city, but should not be confused with the forlorn dying Lion of Lucerne. The Zurich lion is a symbol of courage, strength and power, so often is seen carrying a sword, as in the top corners of this map. On the full cantonal arms, a second lion carries a palm frond, the sign of peace, and the two golden lions (one with sword, one with palm) can still be seen guarding the door of the Zurich *Rathaus*.

The wonderful map portrays Canton Zurich as a lion with the different *Vogteien*, or bailiwicks, coloured in to add to the effect. It is oriented with east at the top so that Lake Zurich can act as the mouth, but that oddly turns the city of Zurich into the uvula, and Greifensee forms one nostril. The eyes were drawn in not only to complete the picture, making Winterthur an unusual eyelid, but also to show that the state was watching you: Big Brother as a lion.

Johann Heinrich Streulin (1661-1742) drew this 1698 map using the famous *Leo Belgicus* maps as an example; the first *Leo Belgicus*, or Belgic Lion, map was drawn by Michael von Aitzing in 1583 and showed the Low Countries in the shape of a lion.

19   The Bernese bear

It's a bear and it's a map. It's a bear map, or the famous *Bärenkarte* of Bern, which is as beautiful as it is political. Here is Canton Bern as it was in 1700, a mighty beast – notice the sharp teeth and sword – that was then the largest canton in the Swiss Confederation.

Legend has it that in 1191 the founder of Bern (Berchtold V, Duke of Zähringen) was hunting in the woods and declared that he would name his new city after the first animal he caught. Luckily it wasn't a rat but a bear. So Bern, with the bear as its emblem, was born.

This is a clever map, both pretty and political. It shows Bern at the heart of its empire, including the 15th-century conquests of Aargau and the Oberland, plus the jewel in the crown: Vaud, taken from Savoy in 1536. From Coppet on the shore of 'Lacus Lemanus' (ie Lake Geneva) to 'Brug' on the River Aare to mountains of the Oberland, it all belonged to Bern.

The map also cunningly omits places which weren't Bernese, such as Biel, then allied to but not part of Switzerland. It doesn't appear at all whereas neighbouring Nidau (ruled by Bern) does, and has its own lake, the 'Nidauer See', now more commonly known as Bielersee or Lake Biel. Today Biel is in Canton Bern, whereas Vaud and Aargau were lost in 1798.

As for Fribourg, that was then a canton but was surrounded by Bernese territory. Rather than give the bear a hole in his middle, the mapmaker simply excluded Fribourg completely. It should be southwest of Bern, near 'Wifflisbourg', or modern-day Avenches (see page 54).

This map from **Jakob Störcklein** isn't the only version of the *Bärenkarte* – another is credited to François-Louis Boizot in 1690 – but is one of the best preserved. It's not big, only 33cm by 23cm, which makes the detailed place names all the more impressive.

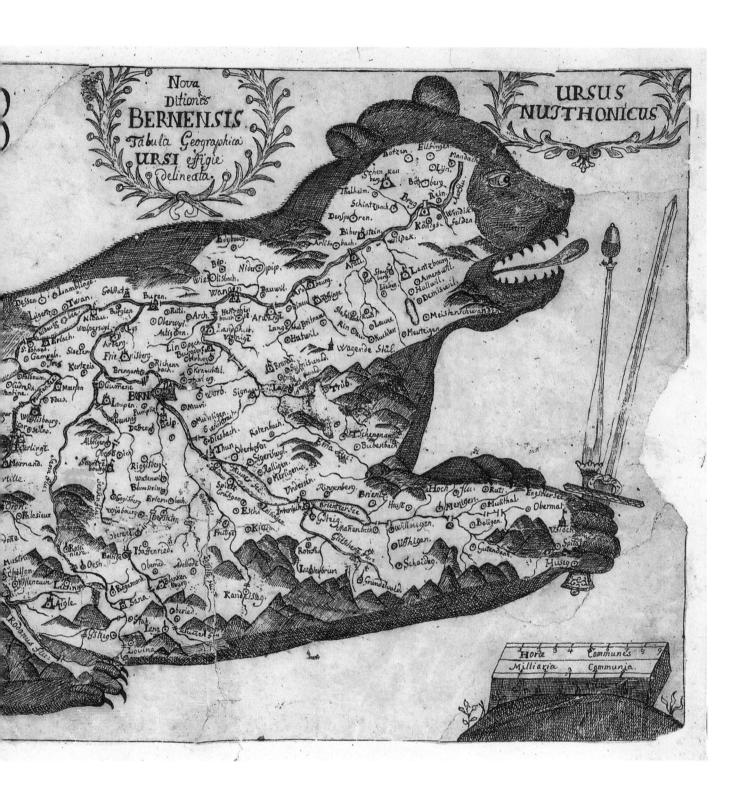

20 A four-way watershed at the heart of Europe

It might look like a map from Lord of the Rings but it is actually a hydrographical depiction of a giant watershed: four rivers flowing in four different directions but each with its source in the Gotthard massif of central Switzerland.

The Rhine is one of Europe's greatest rivers, flowing 1,230km from the Swiss Alps down to the North Sea. Its first 375km are all in Switzerland, making it the longest Swiss river (although not the longest entirely within Switzerland; that award goes to the River Aare). This map shows the Anterior Rhine, recognised as the river's source, flowing north-east to Disentis, and off the map to Chur.

Flowing in the opposite direction is the Rhone, on the left hand side. From its source 'Rhodani Fons' it tumbles down through Valais then (off the map) to Lake Geneva and on into the Mediterranean. Of its total length, the first 264km are in Switzerland, making it the third-longest Swiss river.

In the lakes of St Gotthard near the bottom of the map are the words 'Ticini Fons', ie the source of the River Ticino, which then flows 248km south to join the River Po at Pavia. Today the Ticino's source is recognised as the Nufenen Pass, further to the west. Only the first 91km of the river are Swiss.

The fourth river on this map is the Ursa, although its name has changed. It was originally the Rusa, then for a few centuries the Ursa, until its modern name Reuss came into use. At 158km long, it is the fourth-longest Swiss river, flowing from its source in the Urserental (or 'Vallis Ursaria' on this map) north to Lake Lucerne, and eventually into the Aare.

Johann Jakob Scheuchzer (1672-1733) was a Zurich doctor, mountaineer and cartographer (see page 38). This map first appeared in his book *Ouresiphoites Helveticus* (1723), which detailed his travels through Switzerland in the years from 1702.

A PLAN of GENEVA and the Environs.

PLAN DE GENEVE Ancienne

LAC LEMAN

To Peter Gauſſen
This PLAN is Inscrib'd
Most humble & Oblig'd Serva
I. Rocq

REFERENCES.

1. Bastion de L'ongeinale
2. de St Leger
3. Bourgeis
4. et Hollande
5. du Rhone
6. du Cendrier
7. Place de L'ongeinale
8. Temple St Gervais
9. Place de la Fusterie
10. de Notre Dame
11. de Bel Air
12. de St Gervais
13. Rue de la Pelisserie
14. de la Tour du Boy
15. de la Tartasse
16. de St Antoine ou des belles filles
17. de St Leger
18. Les Prisons
19. La Biblioteque
20. La Discipline
21. La Monnoye
22. La Tour de l'Isle ou de Cesar
23. La Blancherie
24. La Machine Hydraulique
25. Le Nanege
26. L'hostel des Paisibet de Franc
27. La treille
28. Marche du Blé

LE RHONE FLEUVE

L'ARVE

PLAN de GENEVE avec ses Environs.

21

The fortified republic of Geneva

Sandwiched between Savoy, France and Burgundy the independent city-state of Geneva needed to be heavily defended against its bellicose neighbours.

▶

21 The fortified republic of Geneva

◄

Sandwiched between Savoy, France and Burgundy the independent city-state of Geneva needed to be heavily defended against its bellicose neighbours. Its main ally, the Swiss Confederation, was a few days' march away so Geneva built impressive fortifications around its medieval heart.

This map from 1760 shows how the old city was encased in a system of walls and ditches, giving it the appearance of a very spiky hedgehog. It also reveals how close town and country were in those days, with the pastel green patchwork of fields and market gardens reaching right up to the walls.

In December 1602 the city's old defences were almost breached by the Savoyard army, which attacked at 2am on the longest night of the year. Using ladders they tried to scale the walls but hadn't counted on Mère Royaume being up to make soup. She threw her cauldron at the enemy, raising the alarm and saving the city; the event, known as the *Escalade*, is still celebrated every year.

The walls could not keep out the French, who conquered Geneva in 1798 and annexed it as part of the département of Léman, the French name for the lake. With Napoleon's final defeat in 1815, Geneva joined Switzerland as the 22nd canton.

Eventually the walls came down in 1850 to give the growing city more room and make way for modern developments, such as the railway: Geneva's first station opened in 1858 at Cornavin, where the far left spike is on this map. The walls live on in the name Parc des Bastions, the open space that sits where the fortifications once did.

Jean (also John) Rocque (died 1762) fled to England with his parents as French Hugenot refugees. He was later appointed cartographer to the Prince of Wales and is best known for his detailed 24 sheet-map of London, published in 1746.

22 A Prussian principality

From the moment the Swiss broke free from the Habsburgs, they have not been ruled by a king or emperor. With one exception. One canton was for a while both a part of Switzerland and a principality ruled by a foreign king. It was a French-speaking outpost of the Prussian monarchy. It was Neuchâtel.

Named after a castle that was new in the 10th century, Neuchâtel was ruled by the French dynasty of Orléans-Longueville (see page 194). The city's conversion to Protestantism didn't change that situation but in 1707 the death of the last of the royal line, Marie de Nemours, did.

There were 19 candidates to replace her and, as the text at the top of this map states, "Anno 1707 the States have declared the King of Prussia as their sovereign prince". For the electors, it was more important to have a Protestant ruler in distant Berlin than a Catholic one next door in France. Frederick I of Prussia was the new prince.

Apart from briefly being annexed by Napoleon, Neuchâtel remained a Prussian principality for 150 years, despite also becoming the 21st Swiss canton in 1815. It was an odd state of affairs but it worked because Switzerland was then still a loose Confederation rather than one country.

This anomaly lasted until 1 March 1848, when Neuchâtel revolted and declared itself a republic. A few months later Switzerland became a single federal state, with Neuchâtel as an integral part. A royalist putsch in 1856 almost caused a Prussian-Swiss war but peace reigned, even if the Prussian King no longer did over Neuchâtel.

Georg Christoph Kilian (1709-81) was a mapmaker from Augsburg. The map is from an atlas published between 1757 and 1780. Its title reads "Neufchastel or sovereign principality Neuenburg in the Swiss Confederation along with the county of Vallangin".

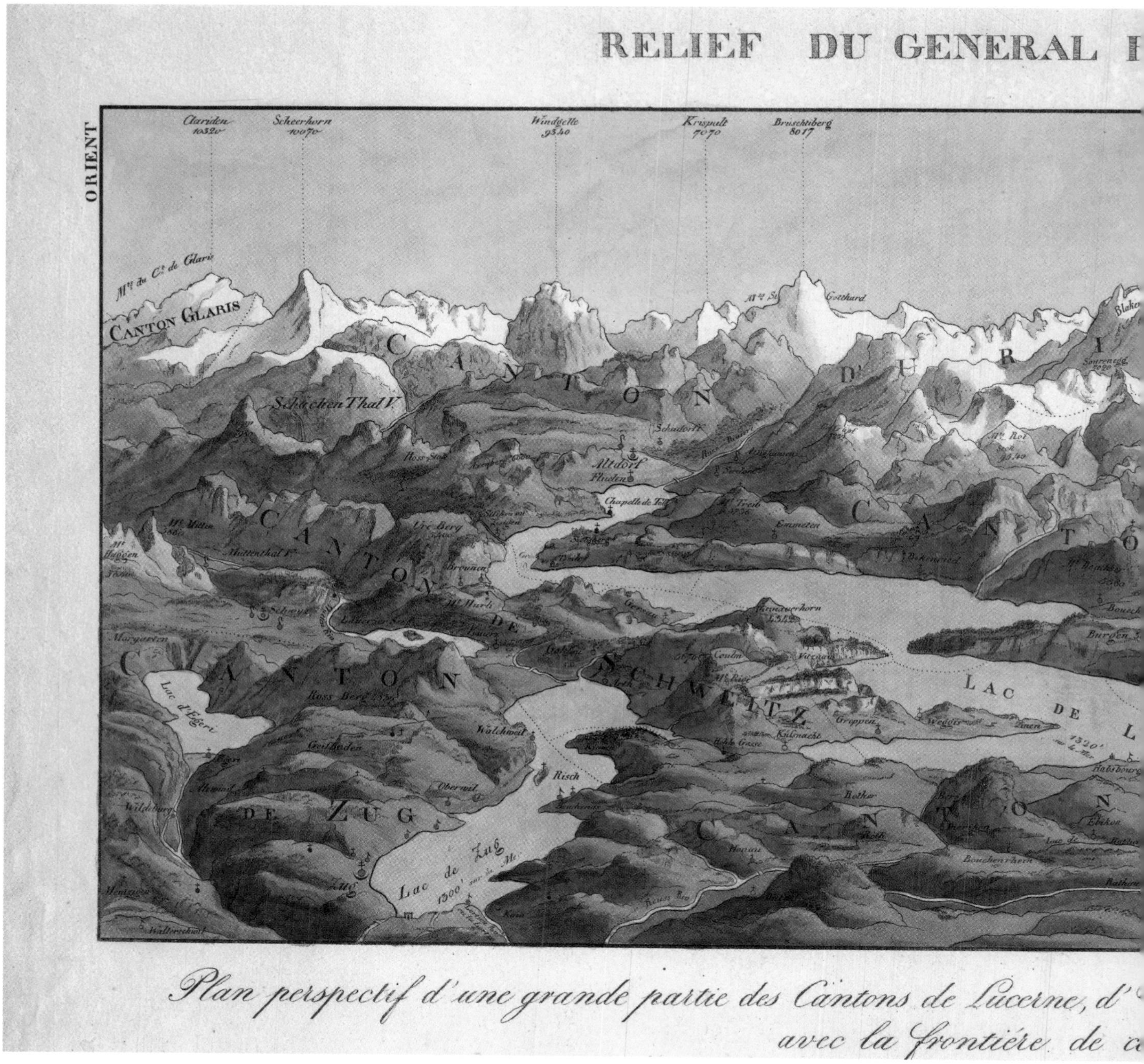

RELIEF DU GENERAL F

ORIENT

Claridon 10320 · Scheerhorn 10070 · Windgelle 9340 · Kraspalt 7070 · Bruschtiberg 8017

Mt. du Ct. de Glaris

CANTON GLARIS

C — A — N — T — O — N

D'URI

Mt. S. Gotthard

Blake

Schachen Thal V.

Schadorf

Altdorf Fluelen

Chapelle de T.

Emmetou

Mt. Mütten

Hüttenthal V.

CANTON DE

SCHWEITZ

LAC DE L

CANTON DE ZUG

Lac d'Egeri

Ross Berg

Risch

Lac de Zug 1300' sur la Mer

CANTON

Plan perspectif d'une grande partie des Cantons de Lucerne, d'

avec la frontière de

23

Lake Lucerne in 3D

It's the world's oldest large relief model and was built from scratch by a Swiss soldier.

▶

23   Lake Lucerne in 3D

◀

It's the world's oldest large relief model and was built from scratch by a Swiss soldier. The Pfyffer Relief of central Switzerland is not only an impressive 3D representation of the Lake Lucerne area, it was a milestone in the development of Swiss cartography.

The model is vast, measuring 6.7 metres by 3.9 metres (so with a total area of 26 square metres), but represents an even larger area of 76km by 44km. In other words: around 3,400km² of the region around Lake Lucerne. It was created from 136 individual pieces so that it could be constructed easily but also taken apart for transport and re-assembly elsewhere.

For its time, it was a very exact representation of the landscape, based on the measurements painstakingly taken by its creator, Franz Ludwig Pfyffer. Its scale is 1:11,500 and although the mountains look slightly higher than they should, it is actually technically accurate. Modern measurements have confirmed how precise Pfyffer's original ones were.

Starting in 1762, Pfyffer took 24 years to survey the land, then build his masterpiece from pulverised brick, wax, cloth, plaster, sand, wire and anything else he could lay his hands on. While the overall model is very impressive, so is the detail of the towns, glaciers, vegetation, streets and streams. It was the first time such detail had been shown in 3D.

Once his relief was finished in 1786, it became so famous that souvenir pictures of it were created, including the one shown here. Since 1873 the relief has been on display in the Gletschergarten Museum in Lucerne.

The map's title reads: "Perspective plan of a large part of the cantons of Lucerne, Uri, Schwyz, Unterwalden, Zug and Glarus and the border of that of Bern"

Franz Ludwig Pfyffer (1716-1802) was from a well-known Lucerne family and served as a lieutenant-general in the French army. This 2D version of his model was made in 1786 as a souvenir for tourists. It measures 61cm × 21cm and is orientated to the south-east.

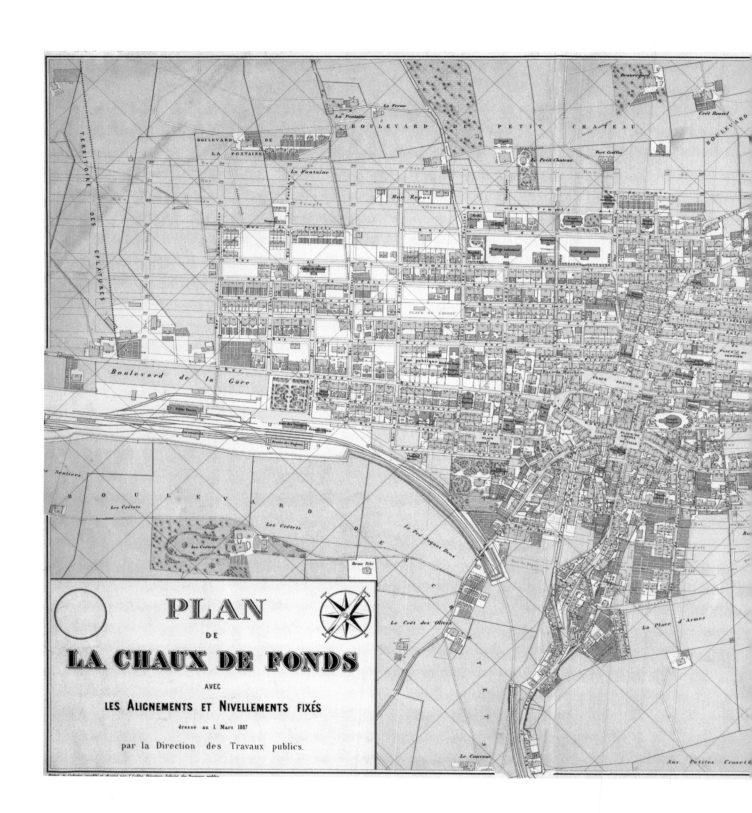

PLAN
DE
LA CHAUX DE FONDS
AVEC
LES ALIGNEMENTS ET NIVELLEMENTS FIXÉS
dressé au 1 Mars 1887
par la Direction des Travaux publics.

24   The perfectly planned town

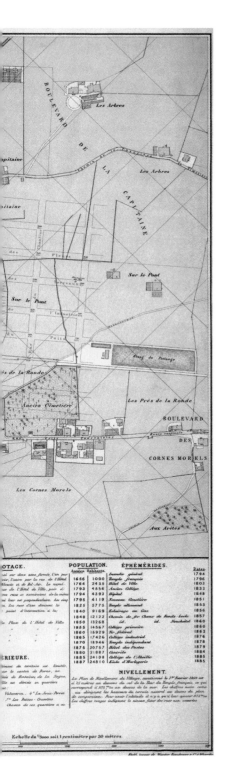

If Switzerland produced only watches, then La Chaux-de-Fonds would be the capital city, not least because it was built with watchmakers in mind. Or re-built, given that it burned down in 1794.

Sitting up in the Jura mountains near France, La Chaux-de-Fonds was incorporated in 1656, when the prince of Neuchâtel gave it the right to have a town hall and weekly market. But it was the development of the watch industry in the 18th century that really put the town on the map, even more once it was almost wiped off the map.

The disastrous fire of 1794 gave planners a chance to rebuild the town with its main industry in mind. It was redesigned on a rectangular grid pattern with rows of identical four-storey buildings arranged so that the watchmakers could get enough sunlight to work but that fires would not spread and snow could be easily cleared.

This map of 1887 shows the grid already in place and developing out past the station, with more streets planned beyond the existing ones. This perfect example of urban planning was given Unesco World Heritage status in 2009, along with nearby Le Locle, with both being cited as "outstanding examples of mono-industrial manufacturing-towns which are well preserved and still active." Karl Marx simply called it a "huge factory-town".

La Chaux-de-Fonds, once the fifth largest town in Switzerland (it is now 13th), is renowned as the birthplace of famous brands such as Omega and famous names like Louis Chevrolet (the car man) and Le Corbusier (the architect). And it's still at the heart of Swiss watchmaking.

An 1887 map from the Office of Public Works, drawn by **J. Lalive** with a northwest orientation. It was printed by Wurster, Randegger & Co in Winterthur, a company of cartographers that was bought by Orell Füssli in 1924.

25   The Italian-speaking canton

Ticino, the southernmost Swiss canton, is a relatively new creation. It joined the Confederation in 1803 as the 18th canton, newly christened after the river that flows from the Nufenen Pass (top left of the map) down into Lake Maggiore. Before that date Ticino had been divided and ruled as a Swiss colony.

Swiss conquest of the southern approaches to the Gotthard Pass (shown at the top of the map) began with Uri annexing the Leventina valley as far as Biasca in the mid-15th century. By 1512 the rest of the canton had been conquered by the Confederation as a whole, and was jointly administered.

That ended with the Helvetic Republic in 1798 (see page 40), which abolished the existing structure and created two mini-cantons of Bellinzona and Lugano. They in turn merged in 1803 to create Ticino – at 2812km², the fifth largest canton in Switzerland. Its highest point is Rheinwaldhorn (Adula in Italian) on the border with Graubünden, 3402m above sea level; its lowest is 193m at Lake Maggiore, also the lowest point in the whole country.

This map from 1812 was the first to show the new canton with its eight districts (and capitals as red dots). For many years the three largest cities – Bellinzona, Locarno and Lugano – took turns at being the cantonal capital, until finally in 1878 Bellinzona became the permanent choice.

The canton is split geographically by the Monte Ceneri Pass, near Lake Maggiore: north is the mountainous Sopraceneri, south is the more populous Sottoceneri. Ticino is the only canton with Italian as its main language, spoken as a first language by 88% of the population.

Heinrich Keller (1778-1862) drew this map in 1812 for *Helvetischer Almanach*, a political periodical that was published from 1799 to 1822. It was based on a map by Paolo Ghiringhelli (1778-1861) and was engraved by Johann Jakob Scheuermann.

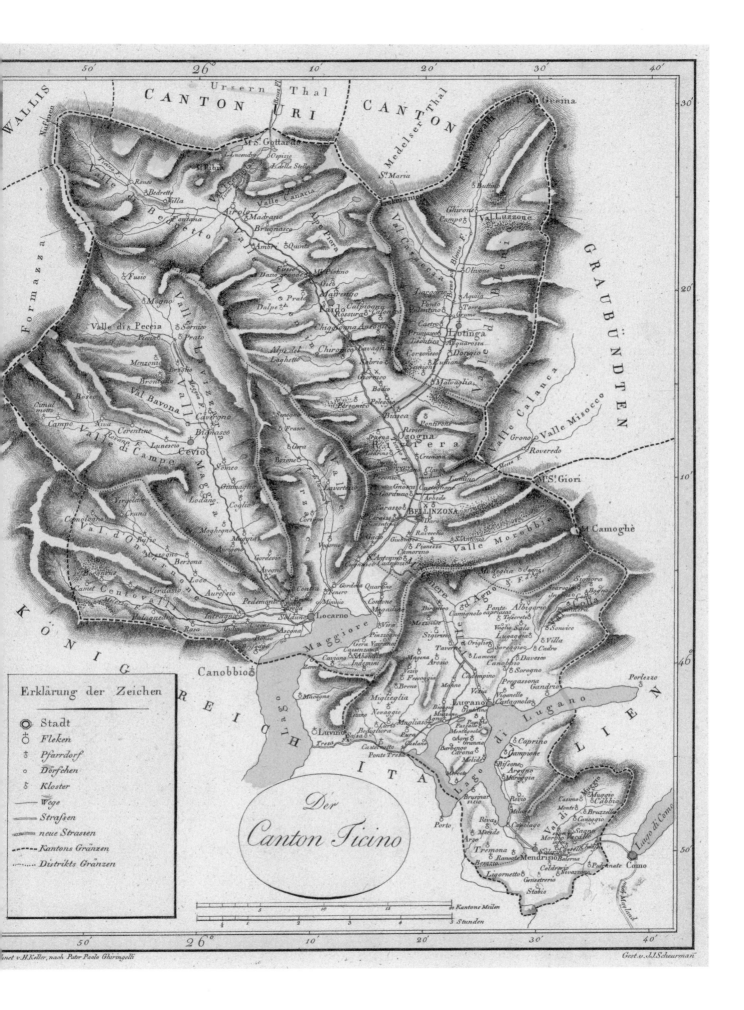

Der
Canton Ticino

26 A tale of two cities

It is the first modern map of one of Switzerland's oldest cities, and one of the first to show the newly unified city. For centuries it had been a medieval version of Cold War Berlin: a city divided in two. One side was Catholic ruled by a man of the cloth, the other Protestant run by men who made cloth. Both were called St. Gallen.

The cities both began with a Benedictine monastery founded in 719 AD but went their separate ways in the Middle Ages. The secular side became a city republic with a Reformed church and trade guilds; the spiritual half remained under a Catholic Prince-Abbott who directly governed large tracts of land. Both prospered from the textile trade and were close (but separate) allies of the Swiss.

In 1566-7 the political separation became a physical one. A nine-metre high *Schiedmauer*, or dividing wall, was built in the heart of St Gallen, completely enclosing the abbey precinct. That was now an enclave within the city republic, which in turn was an enclave within the monastery lands.

After the French invasion, both governments were abolished and the abbey was dissolved. A reunited St. Gallen became the capital of an eponymous new canton in 1803, and the *Schiedmauer* could come down at last, along with the city's actual fortifications. By the time of this map (1828), the dismantling was nearly complete.

Johannes Zuber spent two years using simple surveying equipment to measure distances for the base plan, and then many more months adding in streets and buildings. This was the first map of the whole city to be drawn using modern triangulation techniques.

Johannes Zuber (1773-1853) was born in rural St. Gallen. He was not officially commissioned to create the map but did it voluntarily. The letter codes show city landmarks, such a B for the cathedral and F for the famous library.

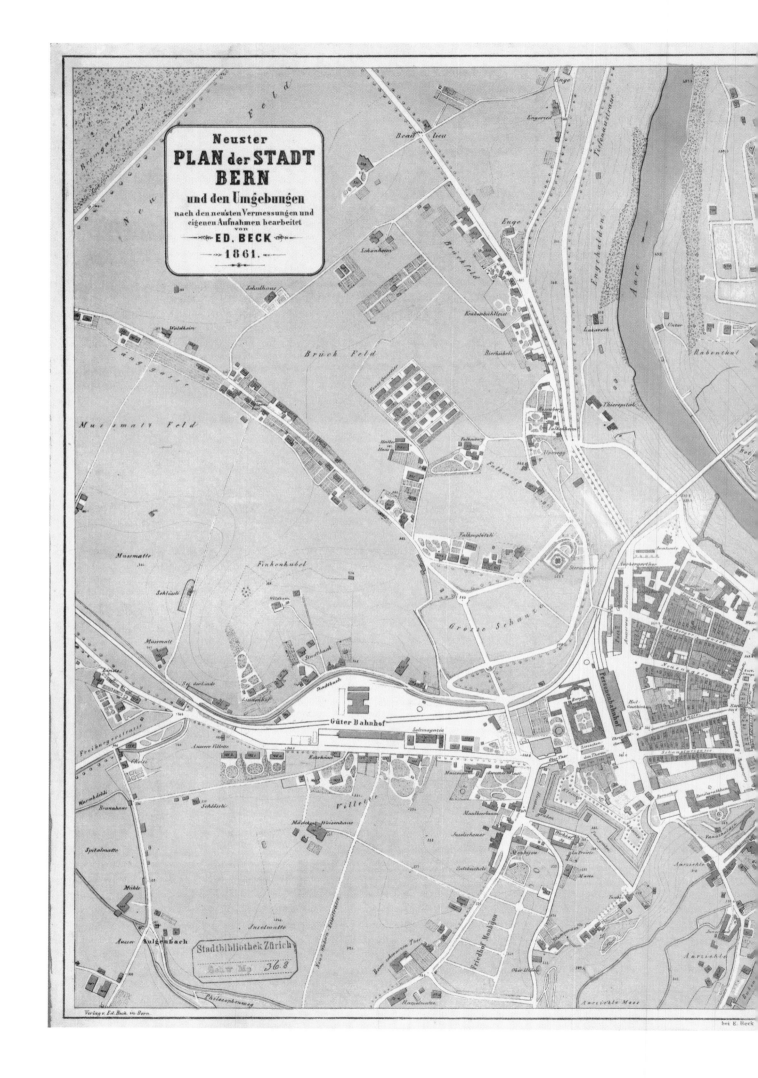

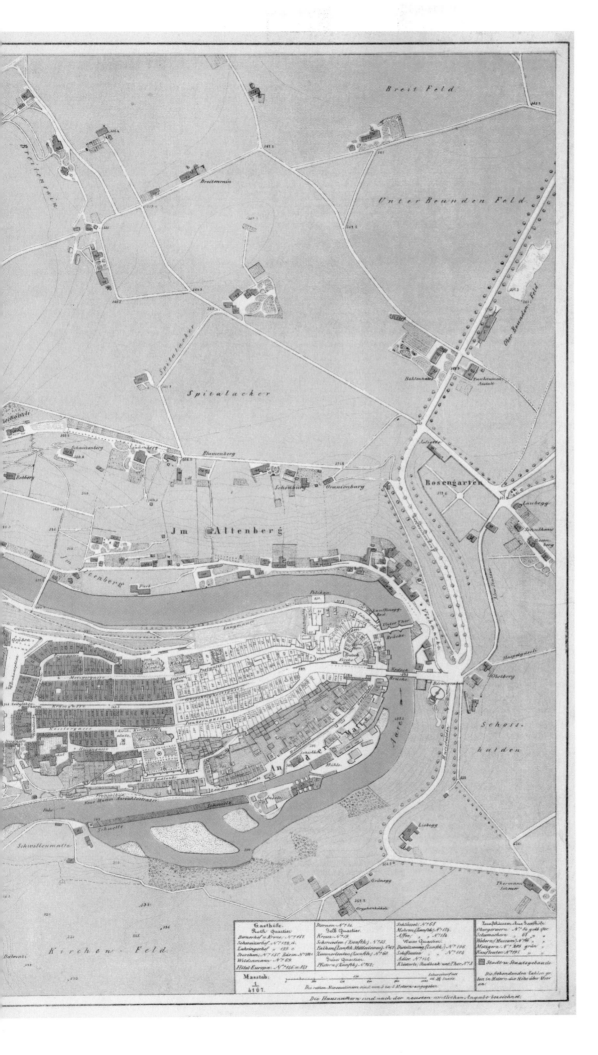

27

The colourful federal city

Officially Switzerland has no capital city but instead a Federal City, or first among equals. Bern was given that status in 1848, prompting it to grow into its new role without quite leaving the past behind.

▶

27 The colourful federal city

◄

Officially Switzerland has no capital city but instead a Federal City, or first among equals. Bern was given that status in 1848, prompting it to grow into its new role without quite leaving the past behind. This map, from 1861, shows Bern with a foot in both the medieval and modern worlds.

The old town in the loop of the River Aare is divided into five coloured zones – red, yellow, green, white and black – a system introduced during the French occupation of 1798. As many French soldiers billeted in the city were illiterate, the street signs were painted different colours so that the men could find the way back to the right part of town for their quarters, even if they had been drinking a bit too much.

Bern's old town still has street signs in the same five colours today. Where the zones meet the signs on either side of the same road are different colours, eg at Kornhausplatz the one on the western side is yellow, on the eastern side green.

That historical hangover is in marked contrast to the changes in Bern at that time. The winged parliament building ('Bundesrathhaus') at the bottom of the red zone had opened in 1857 but two more buildings, including the new domed Bundeshaus (see page 162), would follow to its right, where the 'Casino and Inselspital' are marked in yellow.

As this map shows, the 'Christoffelthurm' still stood in front of the old train station ('Personenbahnhof'). Sadly this huge city gate, which was 55m high complete with a giant statue of St Christopher, was torn down after a referendum in 1864 approved its demolition – by a majority of only four votes.

Eduard Beck (1820-1900) was a German cartographer living in Bern. He states that his map was drawn with "the latest measurements and own recordings" but it has a very odd scale of 1:4 167.

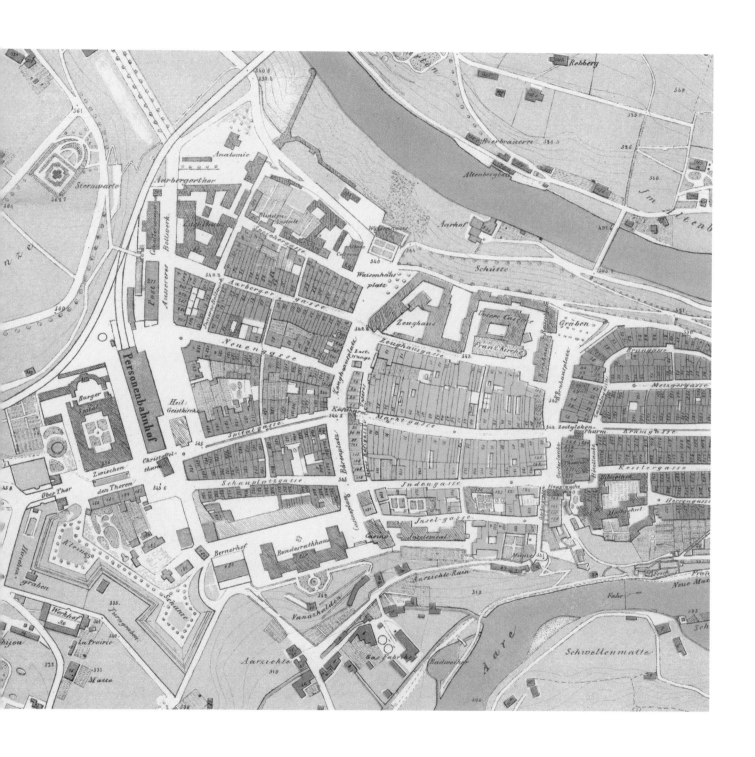

War
&
Peace

Switzerland at the heart
of Europe in conflict

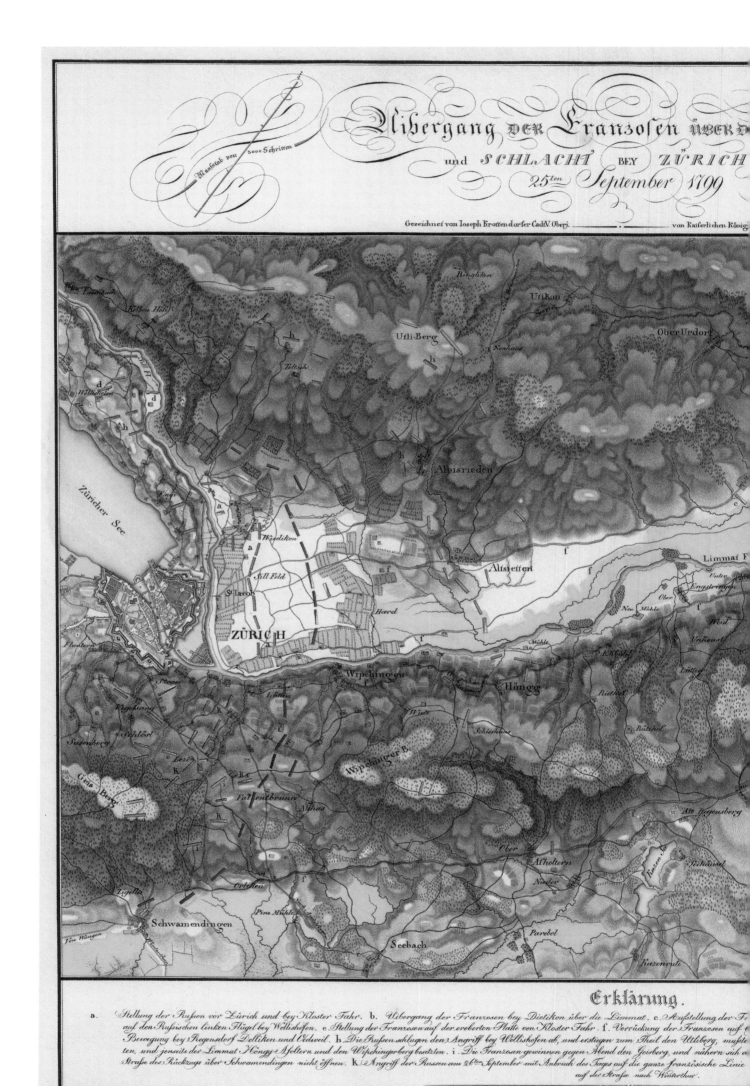

28

Battlefield Switzerland

Neutrality isn't always enough to stop an invasion, as the Swiss discovered in 1798 when the French conquered and occupied the country.

▶

28 Battlefield Switzerland

◀

Neutrality isn't always enough to stop an invasion, as the Swiss discovered in 1798 when the French conquered and occupied the country (see page 40). If that was bad, then there was worse to come when Switzerland became a battlefield for the Great Powers during the War of the Second Coalition.

One crucial clash was the Second Battle of Zurich on 25-26 September 1799, between the French on one side and the Russians (with Austrian support) on the other. The First Battle had taken place three months earlier, resulting in a French defeat, but this time a surprise attack drove the Russians out of Zurich.

This map, oriented with south-southwest at the top, shows the walled city of Zurich (in red on the left) defended by the Russian army (the pale green blocks) in 'Sill Feld' against the French troops (the blue blocks) at Hard; both places are today inner city areas rather than green fields. Downstream at the bulge in the river Limmat at Dietikon is where the surprise French attack at 5 a.m. on 25 September occurred, giving them the upper hand.

The Russian counter-offensives are shown with orange blocks, for example taking Uetliberg (shown with a letter 'h') and also regaining the road to Winterthur on 26 September (shown by the letter 'k'). But the French under Marshal Masséna pushed on to Zurich, forcing the Russians to retreat.

This battle was a vital victory for the French. It was followed by the Russians under Suvorov failing to take the Gotthard Pass, forcing the Coalition to withdraw from Switzerland and prompting Russia to abandon the Coalition.

The text states "Drawn by **Joseph Krottendorfer** CadtV Rifleman in the Imperial-Royal 8th Rifle Battalion". Nothing else is known about him or the map, but its title is very precise: *Crossing of the French over the Limmat and Battle of Zurich on 25th September 1799.*

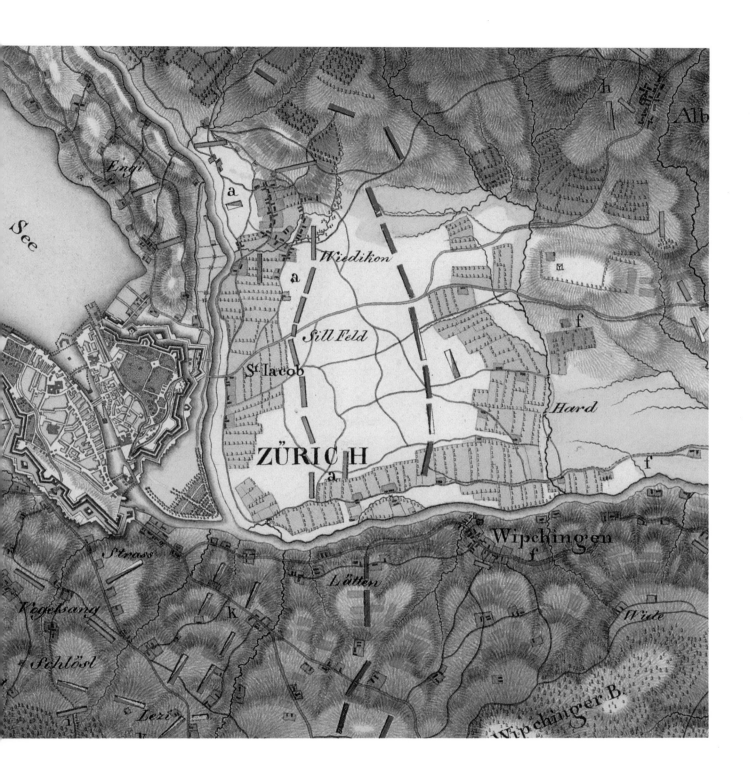

29   The last civil war

On 23 November 1847 a big event took place near the
small town of Gisikon in Canton Lucerne. It was the last
ever battle on Swiss soil, a battle that was the culmina-
tion of a short civil war. Known as the Sonderbund War,
it was the final convulsion of the religious schism
between Catholic and Protestant.

Two years earlier, in December 1845, seven Catholic
cantons – Lucerne, Uri, Zug, Schwyz, Unterwalden,
Fribourg and Valais – had secretly formed a separate
alliance, or *Sonderbund*. Once the other cantons
found out, war was not long in coming, with the
Federal army placed under the command of General
Guillaume-Henri Dufour.

Hostilities broke out on 3 November 1847 when
Sonderbund forces attacked over the Gotthard Pass,
but it was one of few victories for the rebels. On 14
November, Fribourg, geographically isolated from its
allies, was brilliantly captured by Dufour, and Zug capi-
tulated a week later. The crucial canton of Lucerne was
the last bastion for the Catholics.

On 23 November the Federal army won at Gisikon,
and Lucerne surrendered that evening. The remaining
rebel cantons quickly followed, with Valais the last
to surrender on the 29th. The war had lasted 26 days,
resulting in a total of 93 deaths and 510 wounded.
A rather civil civil war.

Of course the war wasn't only about religion,
although that played a role. The conservative Catholic
cantons wanted to protect cantonal rights and
resist the more liberal Protestant cantons' moves
towards a federal state. The upshot was exactly
what the Catholics had fought against: a new constitution and a new federal Switzerland.

This 1847 map by **Franz Malté** was printed in Germany as a supplement for
the *Allgemeine Zeitung*. It shows the seven Sonderbund cantons in brown surrounded by
the Federal cantons; three cantons (Neuchâtel, Basel-Stadt & Appenzell Innerrhoden)
remained neutral.

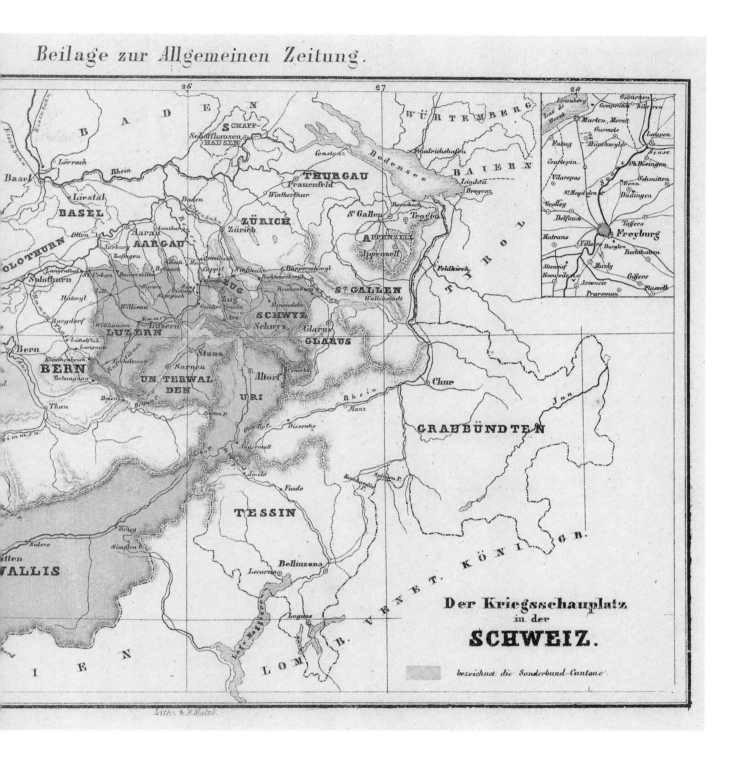

30 Blueprints for a new world

In the aftermath of the First World War the victorious nations wanted to build a better world based on peaceful dialogue. The principal body for this brave new world was the League of Nations, an international organisation created on 10 January 1920.

Even though neutral Switzerland had not taken part in "the war to end all wars", it was a founding member of the new League and an obvious choice for its headquarters. Geneva, already home to the International Red Cross, became the seat of the League of Nations in November 1920. But the League needed a proper home that could house the new organisation.

In 1926, the year Germany was admitted as a member, an architectural competition was launched to design the Palais des Nations, which would be the League's permanent home. There were 377 entries, including one from the famous Swiss architect Le Corbusier, but the jury could not choose a winner. The five leading candidates (Le Corbusier was not one of them) then worked together to draw up the final plans for the building.

On 7 September 1929 the foundation stone was laid in Ariana Park on the edge of Geneva, and the new building opened in February 1936. Three years later war broke out again and the League collapsed, although the building survived. In 1946 the League was replaced by the United Nations, with the Palais des Nations serving at the UN's European HQ since 1966, although Switzerland did not join the UN until 2002.

Post-war extensions were added to the original building, so that the overall complex is now 600m long with 34 conference rooms and 2 800 offices.

The original plans for the Palais des Nations were the result of collaboration from five architects: Carlo Broggi from Italy, Julien Flegenheimer from Switzerland, Camille Lefèvre and Henri-Paul Nénot from France, and Joseph Vago from Hungary.

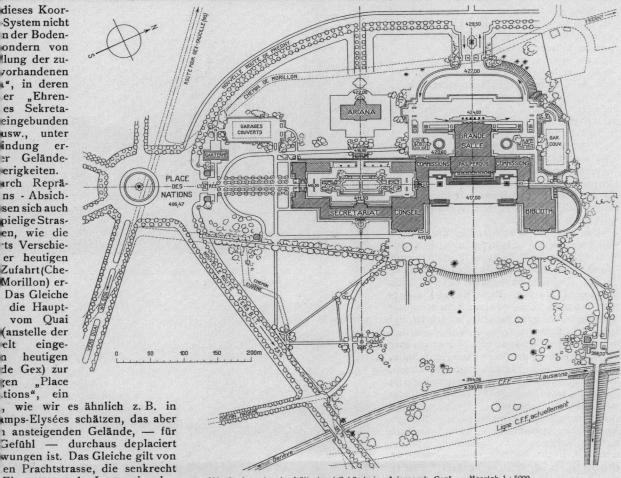

Abb. 2. Lageplan der Völkerbund-Gebäude im Arianapark, Genf. — Massstab 1 : 5000.

dieses Koor-
-System nicht
n der Boden-
ondern von
lung der zu-
vorhandenen
", in deren
er „Ehren-
es Sekreta-
eingebunden
usw., unter
indung er-
r Gelände-
erigkeiten.
rch Reprä-
ns - Absich-
sen sich auch
pielige Stras-
en, wie die
ts Verschie-
er heutigen
Zufahrt (Che-
Morillon) er-
Das Gleiche
die Haupt-
vom Quai
(anstelle der
elt einge-
n heutigen
e Gex) zur
en „Place
tions", ein
, wie wir es ähnlich z. B. in
mps-Elysées schätzen, das aber
anteigenden Gelände, — für
Gefühl — durchaus deplaciert
wungen ist. Das Gleiche gilt von
en Prachtstrasse, die senkrecht
Eingangsaxe des Internationalen
mtes (B. I. T.) am See unten, in
Entfernung ausgerichtet ist.
zige unsymmetrische Linie ist
Allee vor dem Sekretariat, die
rer schönen Bäume wegen bei-
werden muss. —
Wichtigste der uns erteilten
ft ist eine gewisse Beruhigung
lich der Kostenüberschreitung,
Befürchtung auf einem Missver-
zu beruhen scheint; die Sache
sich folgendermassen: Der von
kerbund-Versammlung im Früh-
o bewilligte Kostenvoranschlag
Völkerbundgebäude im Ariana-
lief sich auf 23,63 Mill. Fr.; da-
n für die Bibliothek 4,25 Mill. Fr.,
r, wie schon bemerkt, aus der
ler-Schenkung bestritten wer-
ie Herrichtung der Baustelle
re Entwässerung, hauptsächlich
r Beginn der eigentlichen Bau-
im März 1931 brachte nun die

Abb. 1. Allgemeine Situation (Gebäude mit Front
Richtung Montblanc nicht zutreffend).

me Ueberraschung, dass der *Baugrund im Ariana-*
hr ungeeignet ist. Er besteht aus einer Folge von
- und Lehmschichten, die mit der Tiefe wie auch
ontaler Ausdehnung ganz unregelmässig wechseln.
Feststellung nötigte, nach Massgabe des Fortschrei-
Fundamentaushub, zu vielfacher Aenderung und
ng der Baupläne, woraus die im letzten Sommer
n Verzögerungen infolge „Fehlens der Baupläne"
len sind. Die Fundamenttiefen variieren ganz un-
ssig, bis zu mehreren Metern im gleichen Gebäude-

Es sei erinnert an die Ausführungen Peter Meyers über „Axe
netrie" in Bd. 85, Seite 207 ff. (April/Mai 1925).

teil. Dazu kam das Anschneiden wasser-
führender Schichten, bezw. die Notwen-
digkeit ebenfalls unvorhergesehener um-
fangreicher Entwässerungen.

Alle diese Umstände brachten nicht
nur eine Verzögerung um einige Mo-
nate, sondern auch eine Kostenvermeh-
rung mit sich, die den *Voranschlag auf
26,4 Millionen erhöht* hat, der nun aber
seitens der Bauleitung als endgültig
bezeichnet wird, und von dem am
9. September d. J. der Baukommission
Kenntnis gegeben wurde. Da die Architekten im Ausbau, Material usw. sich
notgedrungen grösste Sparsamkeit auf-
erlegt hatten, wünschte die Baukommission zu wissen, welchen Einfluss
auf die Kosten gewisse reichere Aus-
führungsweisen, opulentere Umgebungs-
arbeiten und dergleichen haben *würden*.
Für diese „bessere Ausführung" nannten
die Architekten am 17. September den
Betrag von 29,9 Millionen, bezw. von 5,4 Millionen für
die Bibliothek. *Diese* Summen differieren also mit den
für 1930 genannten um die bewussten rund 7 Millionen
sogenannter „Ueberschreitung". Es handelt sich also
dabei wie gesagt nur um eine informatorische Berechnung;
die durch die Fundationsschwierigkeiten im Arianapark
verursachte, wirkliche Ueberschreitung beträgt 26,4—23,63
= 2,77 Mill. Fr.

Uebrigens sei bemerkt, dass die Verlegung der Bauten
vom See weg in den Arianapark im Herbst 1928 ohne
Befragung der Architekten vom Völkerbund beschlossen
worden ist. Das Arianagelände war, im Gegensatz zur

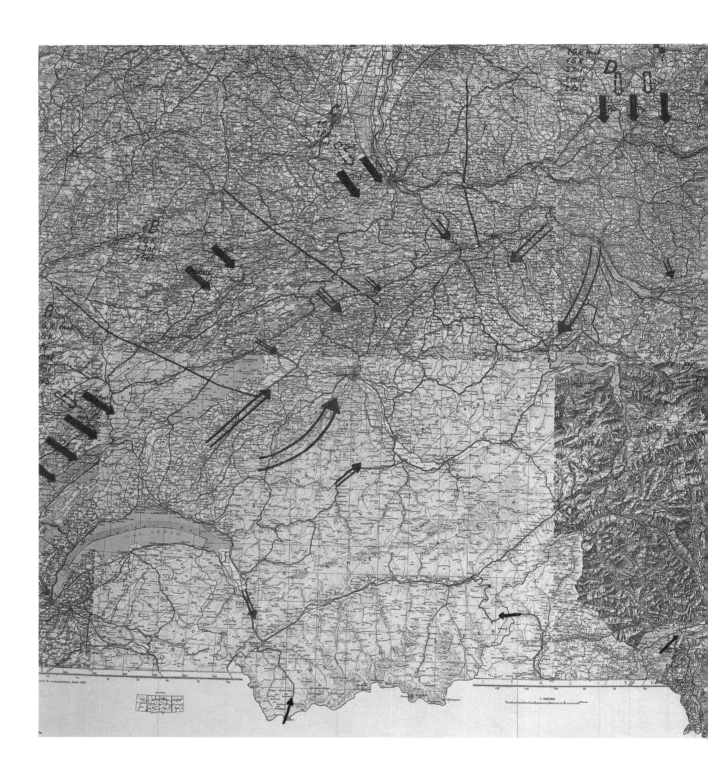

31 The invasion that never came

In May 1940 it seemed to be a matter of when not if Germany would invade Switzerland. Armed neutrality had not saved Belgium, and a German flanking manoeuvre around the southern end of the French Maginot Line was highly likely. Switzerland would be the back door to a German victory over France.

The Swiss braced themselves for invasion and, with their backs to the Alps, declared a policy of no retreat and no surrender. They would fight until the bitter end, using steel when no bullets were left. It didn't prove necessary in May but the French surrender in June prompted the next crisis.

With France defeated and Italy in the war, Switzerland was surrounded by the Axis powers, who developed various strategies for dealing with the Swiss. The first ideas of invasion came on 25 June, with pincer attacks by Nazi forces in France, Austria and Germany, joined by one from Italy. In this scenario a victorious Germany would occupy most of the country, while Italy would get Valais, Ticino and Graubünden.

The military aim was to gain control of the industrial lowlands while preventing a Swiss retreat into the mountains, where the army could fight on from its defensive redoubt (see page 108). Crucially the Alpine rail tunnels, vital transport links between Germany and Italy, had to be captured before they were destroyed.

The German plans, revised in August and again in October, became known as Operation Tannenbaum, although they were just one of many different strategies at that time. But the invasion never happened and Switzerland escaped.

An Operation Tannenbaum map (October 1940) from the German military archives. The Germans would invade from north, east and west (blue arrows) with Italians from the south (smaller black arrows). Bold arrows show the first wave, hollow ones the second.

32   Housing the wartime guests

The defeat of France in 1940 gave neutral Switzerland a new challenge: retreating soldiers, and not just a couple but the 45th Battalion of the French Army. On 20 June over 42,000 soldiers – and their 5,500 horses – crossed the border into Jura, two days before that border closed.

At first the soldiers were housed in schools, gyms and barns, then in camps all across Switzerland. Most of the French were repatriated once the fighting in France stopped, but some stayed, as did the 12,152 Polish soldiers who had fought beside the French. The Poles presented a particular problem as Poland no longer existed, so they had to stay until the end of the war.

In July the Swiss built a Polish "concentration camp" near Büren an der Aare, literally to concentrate them in one place. The camp had 120 wooden barracks with one watchtower, as much to prevent the Poles escaping back to the war as to protect the local women from international relations. Neither was entirely successful.

This map shows where foreign soldiers were in northwest Switzerland, the main area for internees. Coloured dots denote the nationalities (red for French, blue for Polish and one yellow one for the 86 Brits) with one special case. The red dots in the 'Mentue' district (beside Lake Neuchâtel) are for the Spahis, the Moroccan and Algerian soldiers of the French Army kept separate from the rest.

More than 100,000 foreign soldiers were interned in Switzerland during the war, some for the whole duration. Many worked on farms and in the mountains, building roads and clearing woods.

A map from 16 December 1940 with the location and numbers of foreign military internees. For example, in District 6 of Seeland Nord were 6 French and 6 Polish officers, 528 French and 593 Polish soldiers, plus 82 French and 320 Polish horses.

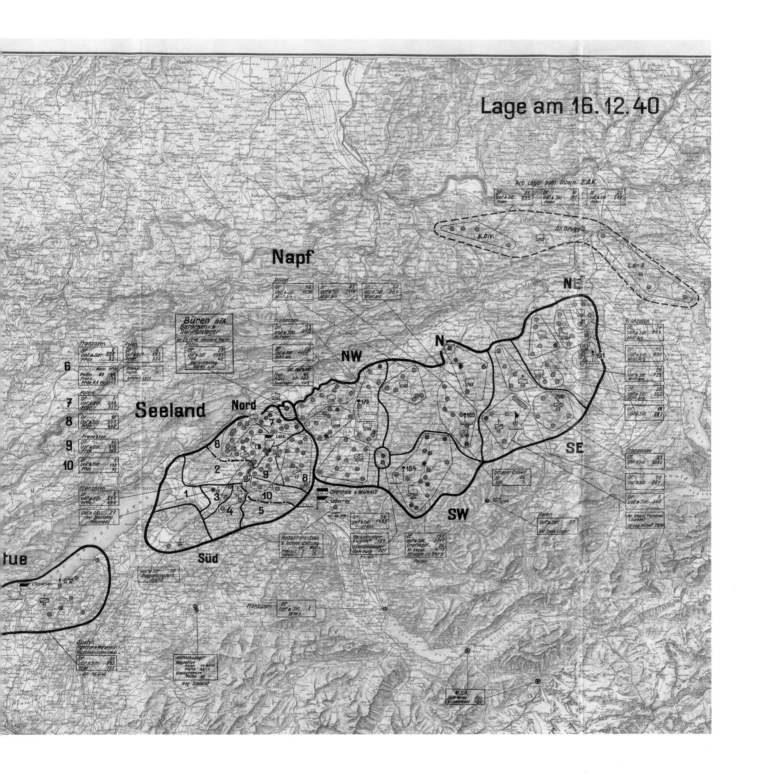

Lage am 16.12.40

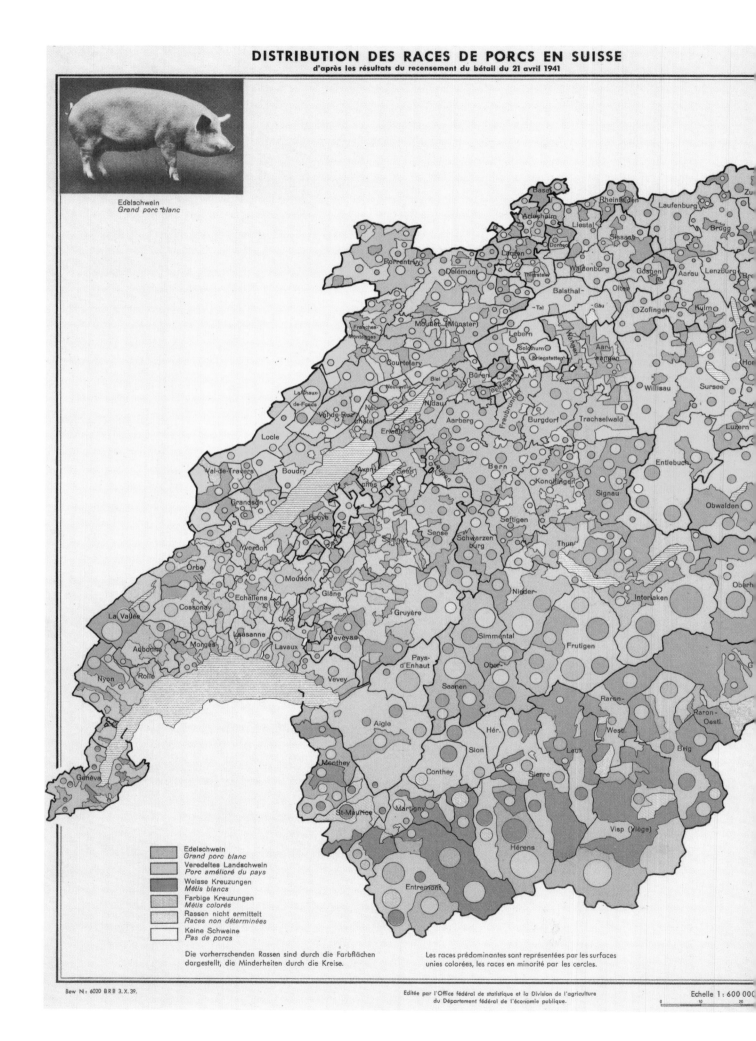

DISTRIBUTION DES RACES DE PORCS EN SUISSE
d'après les résultats du recensement du bétail du 21 avril 1941

Edelschwein
Grand porc blanc

Legend:

- Edelschwein
  Grand porc blanc
- Veredeltes Landschwein
  Porc amélioré du pays
- Weisse Kreuzungen
  Métis blancs
- Farbige Kreuzungen
  Métis colorés
- Rassen nicht ermittelt
  Races non déterminées
- Keine Schweine
  Pas de porcs

Die vorherrschenden Rassen sind durch die Farbflächen
dargestellt, die Minderheiten durch die Kreise.

Les races prédominantes sont représentées par les surfaces
unies colorées, les races en minorité par les cercles.

Bew N: 6020 BRB 3.X.39.

Editée par l'Office fédéral de statistique et la Division de l'agriculture
du Département fédéral de l'économie publique.

Echelle 1: 600 000

ITUNG DER SCHWEINERASSEN IN DER SCHWEIZ
nach den Ergebnissen der eidg. Viehzählung vom 21. April 1941

Veredeltes Landschwein
Porc amélioré du pays

Schweinedichte
Densité de l'exploitation porcine

Anzahl Schweine pro km² Kulturfläche
Nombre de porcs par km² de surface cultivée

Herausgegeben vom eidgenössischen Statistischen Amt und von der
Abteilung für Landwirtschaft im eidgenössischen Volkswirtschaftsdepartement

Kümmerly & Frey, Bern

33

A wartime pig census

It's 1941 and while the rest of the world is busy counting bodies, the Swiss are counting their pigs.

▶

33 A wartime pig census

◄

It's 1941 and while the rest of the world is busy counting bodies, the Swiss are counting their pigs. And cows and goats. And then making maps to show where all the animals are. That's not as daft as it sounds, given that neutral Switzerland was surrounded by the Axis powers. Food self-sufficiency was crucial to surviving the war so a livestock check was literally vital. The maps maybe weren't essential but the data behind them was.

This spotted pig map shows where the different breeds were found, with districts coloured according to the main breed, and minority breeds in the circles. At the top are photos of the two major breeds: the *Edelschwein* (left-hand picture and in red on the map), which was largely in Canton Bern, and the predominant *Veredeltes Landschwein* (right and in yellow) that made up 45% of all pigs.

The inset map shows density of pigs per km². Thurgau and Appenzell were clearly the cantons with the densest pig population, for example in Thurgau it was 111 pigs per km², compared to a national figure of 33 – and only 6 per km² in Graubünden.

At that time the total pig population was 764,378, the lowest number since the mid-1920s. They were owned by 154,726 farmers, with two thirds only having one or two pigs. It was not meat production on today's scale, as only 10% of farmers had more than 50 pigs. In fact just ten farms had more than 500 pigs.

Today Switzerland has 1.6 million pigs across 8,300 farms. Canton Lucerne is now the porcine capital, with more pigs than people.

The map was published by the Federal Statistics Office in 1944 as part of a series using the livestock census of 21 April 1941. The canton with the largest pig population was easily Bern (20% of the total), followed by Thurgau and Vaud (both 9%).

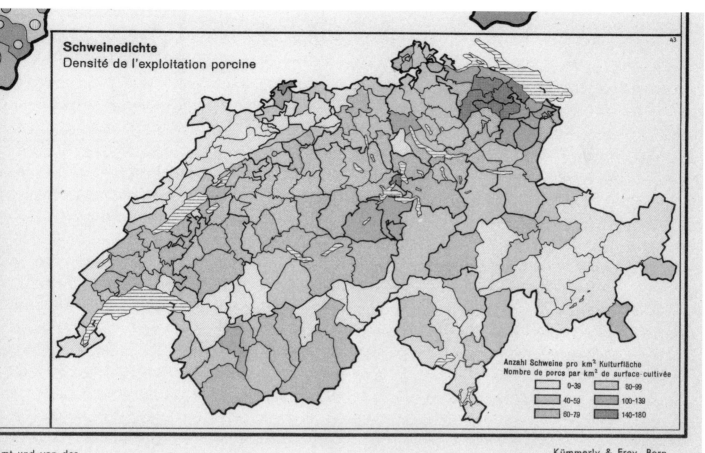

Schweinedichte
Densité de l'exploitation porcine

43

Anzahl Schweine pro km² Kulturfläche
Nombre de porcs par km² de surface cultivée

0–39		80–99
40–59		100–139
60–79		140–180

Amt und von der
irtschaftsdepartement

Kümmerly & Frey, Bern

34   The national redoubt

An imminent German invasion in 1940 (see page 100)
led to a radical rethinking of Switzerland's defensive
position. Out went the traditional heavy defence
at the border, in came a fortified redoubt in the mountains. It was total defence in the face of total war.
It was Fortress Switzerland.

Switzerland never has a general in charge of the army
during peacetime but elects one when needed. The man
chosen in August 1939 was Henri Guisan from Vaud,
who declared in 1940 that there would be no surrender;
the Swiss would fight to the last man, if necessary from
highly fortified positions in the Alps.

Guisan's redoubt was militarily sound - using the
natural fortress of the mountains - but politically risky
as it meant abandoning all the main Swiss cities and
industry, along with 80% of the population. But he had
few other options. The nation of 4.2 million people,
whose president was talking of accommodating the new
European order, was totally surrounded by the Axis
powers.

Come the invasion, frontier troops would delay
the enemy long enough for the main army to reach the
réduit, or redoubt. Stocked with enough food and
ammunition for months, they could prevent the total
conquest of Switzerland. It was as much deterrence as
defence: invade us and we will blow up our tunnels
and tie you down in the mountains until the very end.

Guisan's redoubt was never put to the test as the
invasion never came. Historians are divided on how
effective it would have been, but perhaps its existence
in theory was enough for it to succeed in practice.

The Swiss military situation on 24 May 1941 with four operational areas. The redoubt's
defensive lines are in red, spanning the Alps from Fortress Sargans on the Austrian border
across the Gotthard fortifications to St Maurice in the western Rhone valley.

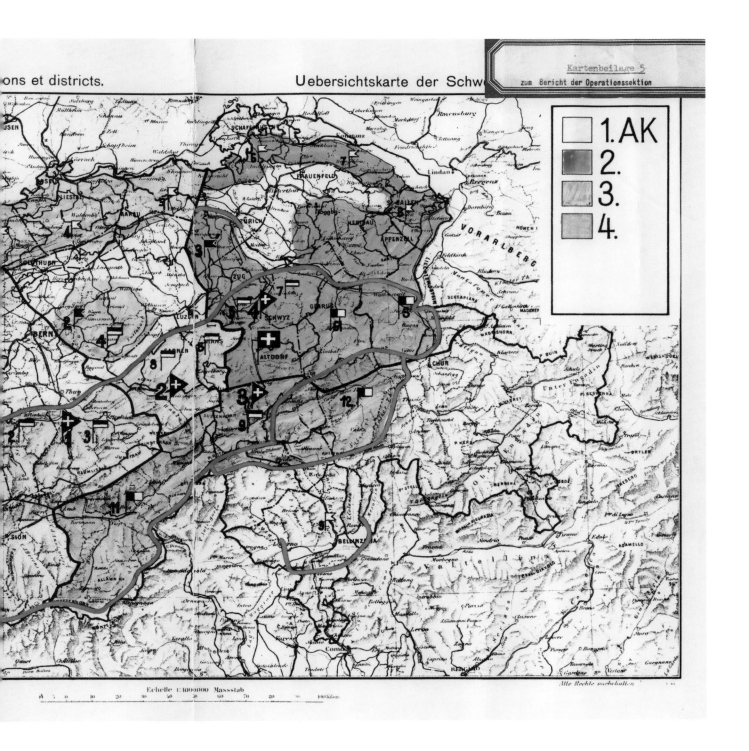

35   A silk lifesaver

An Allied pilot shot down over enemy lines needs a map to escape, especially if he is to reach neutral Switzerland. Paper can tear, is bulky and hard to hide, and doesn't like water; most crucially it rustles when opened, giving away a hiding position. So during the Second World War British RAF pilots were issued with a very special escape map.

In 1939 the British War Office created MI9, an intelligence department specifically designed to help servicemen in enemy territory and assist prisoners of war (POWs). The philosophy of evading capture and escaping to safety was paramount to MI9's ethos, but it needed the right maps. They had to be thin, light, resilient, water-resistant, easily concealed and silent to open.

One MI9 officer, Christopher Clayton Hutton, is credited with perfecting the escape map by using parachute silk. His first trial at printing on silk was hopeless as the ink ran and blurred, but he solved that by adding pectin. Just as pectin sets jam, so it also set the ink to the silk, making maps clear and legible even when very detailed.

Over 1.75 million copies of more than 250 different maps were produced by the British War Office during the war. Not only on silk but also on man-made fibres or tissue paper made from mulberry leaves. The edges were sewn or glued (as with this map) to stop fraying.

Material maps were included in escape kits and could double up as a bandage or a hanky. Many were also smuggled into POW camps, for example hidden flat within the board of a Monopoly set.

Escape map 43/D produced in 1943 by the British War Office; it measures 87cm by 74cm and is printed on both sides in eight colours. It was never used. The green line near Geneva marks the border between Nazi Occupied France and Vichy France.

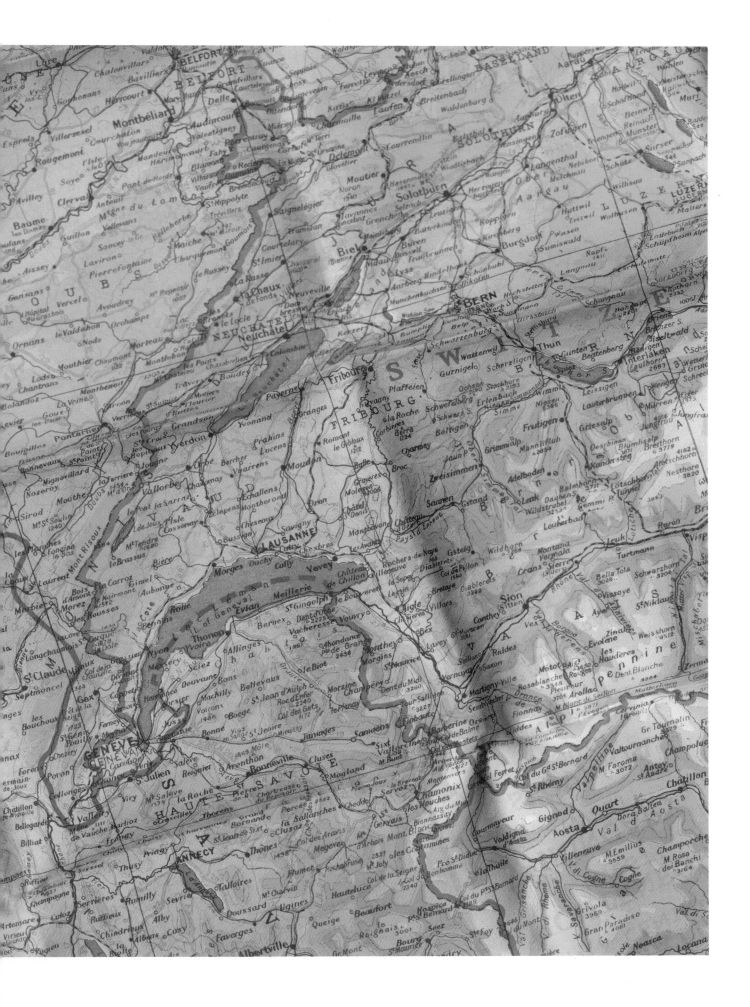

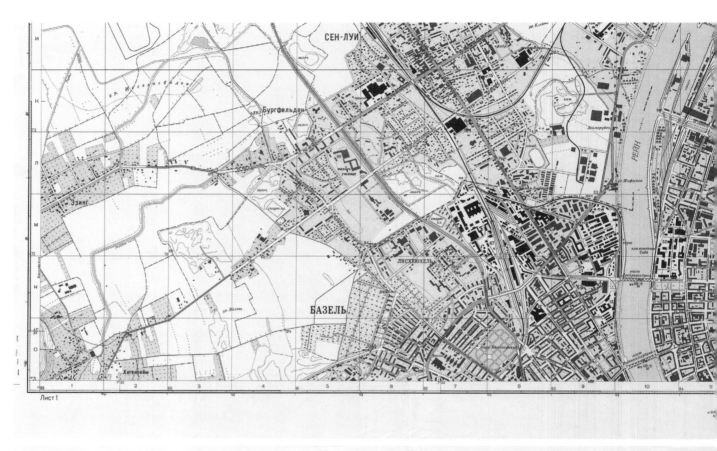

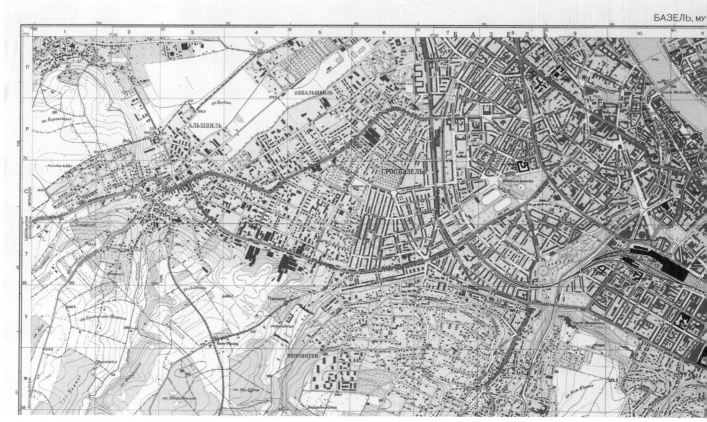

36

The Soviet army in Basel

It's not hard to imagine the scenario:
1975 and Soviet tanks rumble across into
the west.

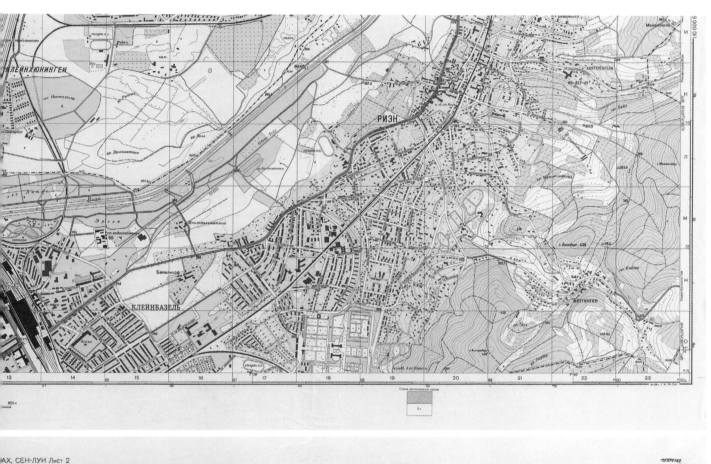

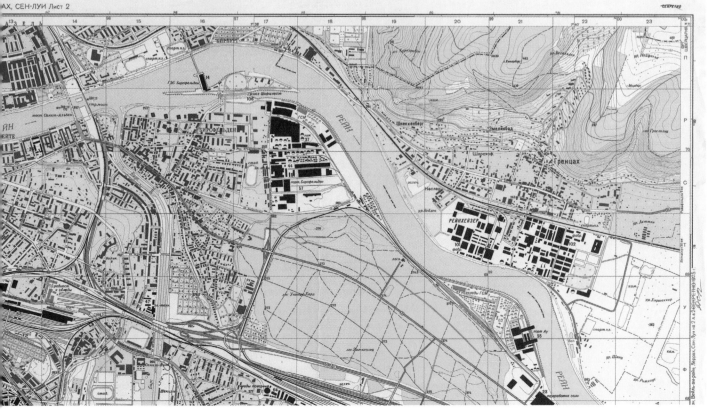

36   The Soviet army in Basel

◄

It's not hard to imagine the scenario: 1975 and Soviet tanks rumble across into the west. Nato defences are taken by surprise and before long the Red Army is at the Rhine and about to invade Switzerland. And an invading army needs maps.

The Soviet general staff mapped out the whole of Europe, adapting what was available in each country to their own needs. City maps were usually at a scale of 1:10,000 to give enough detail for invading and occupying troops. In Switzerland's case that meant the likes of Bern, Geneva, Zurich, St. Gallen, Lucerne – and Basel, as shown here across two maps. Or, Базель as it is written in Cyrillic.

Important locations are colour-coded and numbered. Purple is for government buildings, so 60 is the *Rathaus* (or town hall), 76 the courthouse, 95 the university and over the river in Kleinbasel, 12 is the labour exchange. Each bank (Банк) is also shown, eg numbers 2 to 7 in the inset map.

Industrial targets are in black, such as the principal train stations (72 and 73) or chemical factories (50 and 52), while communication targets are green, for example 58 is the central post office. All would have been primary military objectives so needed to be clearly identified.

Each bridge across the Рейн (or Rhine) has detailed data on its size and strength. So the *Mittlere Brücke*, at the top of the lower map, stands at 254m above sea level and 10 m above the river. It is 180m long and 17m wide and can carry up to 60 tonnes in weight. That's crucial information for tanks wanting to cross the Rhine into Switzerland.

Two Soviet military maps of Basel from 1975. The placenames in the title have been transcribed into Cyrillic so Russians could pronounce them more accurately. So СЕН-ЛУИ and РИЭН are literally 'Sen-Lui' and 'Rien' or Saint-Louis and Riehen.

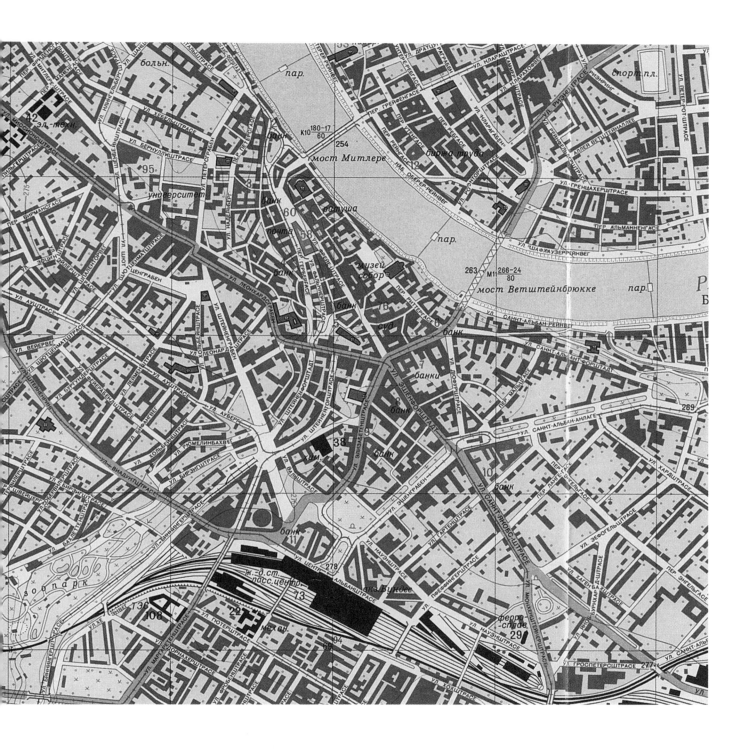

Transport & Tourism

The development of trains, planes and automobiles and the tourists who came with them

37 Travelling with horse power

Before the age of the steam train, public transport was simply a coach and horses. As journeys were uncomforable it was essential to know the distances in order to anticipate the length of suffering involved. This *Postkarte* (literally "post map") of 1799 shows the main Swiss coach routes.

Distances are given in "posts", a set measurement equal to two German miles, also known as geographic miles. A German mile was 1/15th of a degree of longitude, or roughly 7,420 metres, so one post was equivalent to twice that, ie 14.84 kilometres (though the metric system was only introduced in the same year as this map). Each line is marked with dots and dashes, representing fractions of a post.

The key to this cartographical Morse code is in the top left. A dot was a quarter post, a short dash not crossing the line was a half post, a short dash and a dot was ¾ of a post, a long dash crossing the line was one post, and so on. Thus the route from Sempione (or Simplon) to Divedro in Italy is two long dashes and one short, or 2½ posts, ie 37.1km.

Lines are shown without exact relation to geography or the actual route. They are representations, much like the London Underground map, that simplify the map to help the reader but create some unlikely direct connections, eg in central Switzerland, from Amsteg to Disentis.

This Swiss *Postkarte* was number 19 in a series of 40 similar ones that were published weekly, starting in January 1799. By October the whole atlas would have been complete, including 14 sectional maps of the Holy Roman Empire, six of France and four of Britain.

Franz Johann Joseph von Reilly (1766-1820) was an Austrian cartographer who published several atlases, including the *Allgemeiner Post-Atlas von der ganzen Welt*. It didn't cover the whole world but had this map of Switzerland as it was then, without Graubünden.

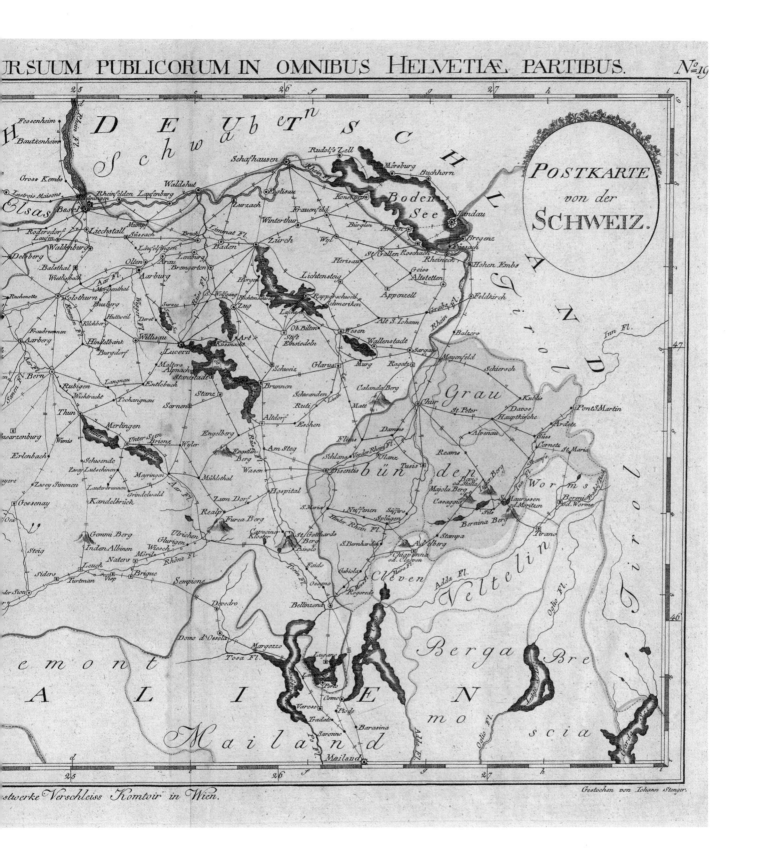

RSUUM PUBLICORUM IN OMNIBUS HELVETIÆ PARTIBUS.   Nᵒ 19

POSTKARTE
von der
SCHWEIZ.

stwerke Verschleiss Komtoir in Wien.   Gestochen von Iohann Stenger.

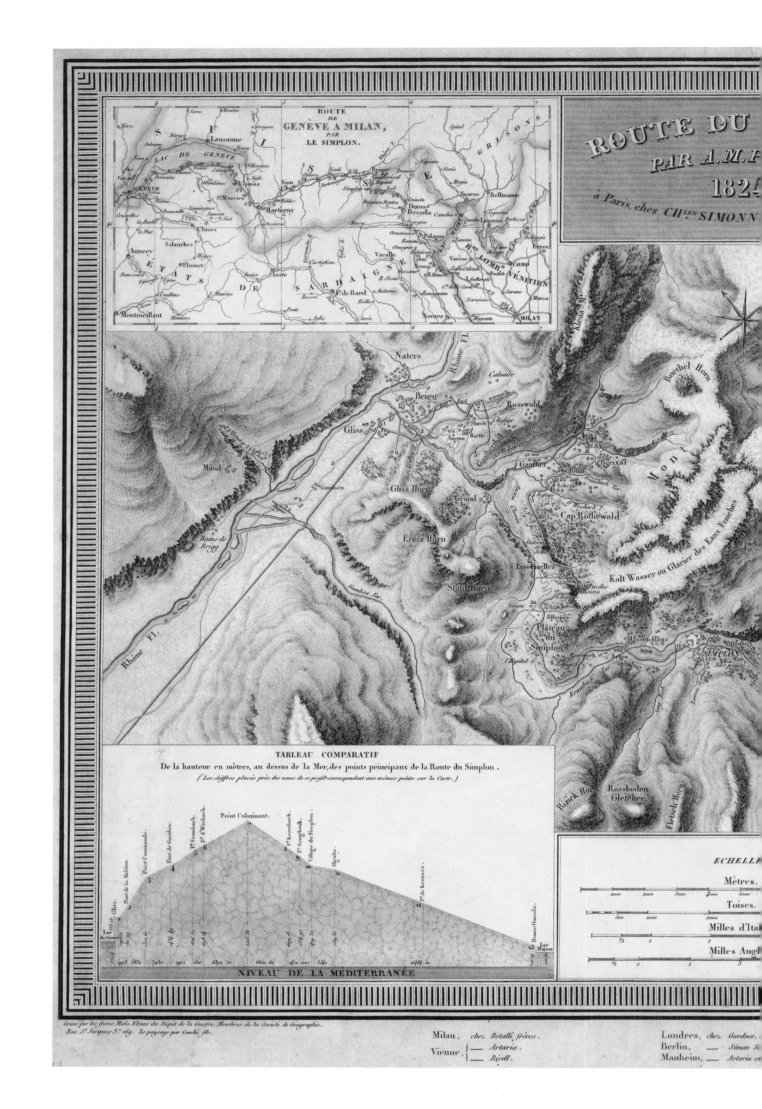

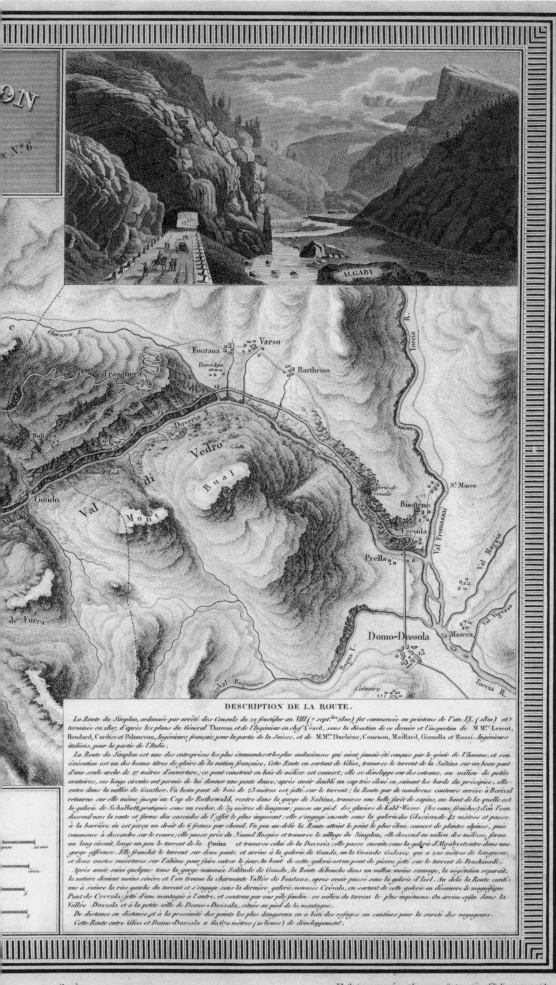

38

Over the Simplon Pass

It was the first modern road over the Alps, a marvel of transport engineering built as a military route by Napoleon.

▶

38   Over the Simplon Pass

◄

It was the first modern road over the Alps, a marvel of transport engineering built as a military route by Napoleon. At its highest point, 2005 m above sea level, it crosses the Simplon Pass that separates the Rhone and Po valleys.

The road over the Simplon has had a rocky history. Once an important trade route, it was bypassed in favour of the Gotthard to the east but when the latter's trade was crippled by the Thirty Years' War, the safer Simplon thrived again. The road was rebuilt by the Valais salt trader Kaspar Stockalper, but his fall from power in 1678 meant that the Simplon also lost its way. Then came Napoleon.

Monsieur Bonaparte wanted a new road wide and safe enough for his artillery to reach Italy quickly. In 1801 he appointed Nicolas Céard as chief engineer and construction began that spring. As the inset picture shows (top right), the completed road was eight metres wide with galleries and tunnels as protection. Napoleon never actually used the road but thousands of others did.

Its solid construction made it the fastest coach route over the Alps, at least until the coming of the train. In 1906 the Simplon rail tunnel – then the longest in the world – opened and the road was used less and less. The national road plans of the 1950s (see page 142) upgraded the Simplon to be accessible all year round, making it a vital trade link once again.

The small map top left shows the whole road between Geneva and Milan, via the Simplon Pass. It goes through three countries on the way: Switzerland and two kingdoms to the south, Sardinia (including both Savoy and Piedmont) and Lombardy-Venetia.

A map from 1824 by **Aristide Michel Perrot** (1793-1879) and published by Charles Simmoneau of Paris. It shows the 60.7km stretch of road over the Simplon Pass, between 'Gliss' in Valais and 'Domo-Dossola' in Italy, built between 1801 and 1805.

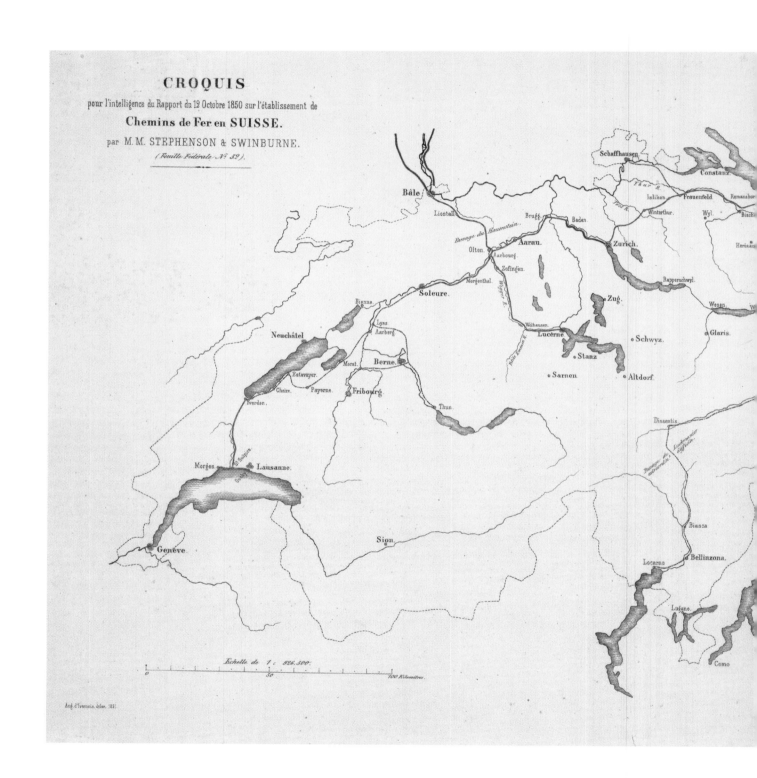

CROQUIS

pour l'intelligence du Rapport du 12 Octobre 1850 sur l'établissement de

Chemins de Fer en SUISSE.

par M.M. STEPHENSON & SWINBURNE.

(Feuille Fédérale Nᵒ 52).

39   The British rail plan for Switzerland

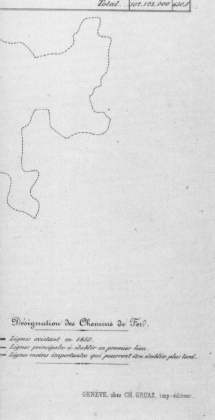

Railways are an iconic feature of Switzerland, making the Swiss network one of the most well-known and most well-used in the world. But it wasn't always that way.

1850 Switzerland had less than 30km of railways: just one line from Zurich to Baden, known as the Spanisch-Brotli-Bahn. In contrast Britain already had 10,000km of lines, thanks to George Stephenson and his revolutionary steam train, the Rocket. The new Swiss federal government was only two years old but it decided it was time Switzerland had a national network.

The government invited Stephenson's son Robert to come to Switzerland and map out exactly such a network. With Henry Swinburne, he assessed the hills and valleys, and devised a national rail plan: a cruciform network of two main lines crossing at Olten, the west-east axis running between Lakes Geneva and Constance then on to Chur, the north-south one from Basel to Lucerne. The Alps were seen as impassable.

A note on the map shows that he planned 650km of railways at a cost of 102 million French francs (the Swiss franc was not yet in circulation), or about £4 million at that time (about 1,500 million Swiss francs today). A lot of money for what was then a poor country.

The plan was not adopted by the Swiss, who preferred a cantonal solution with private financing, but in the end the network grew roughly along the lines mapped out by Stephenson. His influence on Swiss railways can be seen today: the trains run on the left, just like in Britain, and Olten is still a major crossing point at the centre of the network.

The Stephenson and Swinburne railway report was presented in French on 16 November 1850 as *Feuille fédérale Suisse No 52* (Year 2, Volume 3); in German it was on 23 November as *Schweizerisches Bundesblatt No 53*.

40 Mapping the panorama from Rigi

For Victorian visitors to Switzerland there was one must-see sight that was top of every list: experiencing the 360° panorama from the summit of Mount Rigi, preferably at sunrise. So what better souvenir that a circular map that cleverly merges the panoramic view with a normal map.

At the centre is Rigi, known as Queen of the Mountains and sitting regally surrounded by lakes and peaks. From Lucerne it was a short boat ride across the lake to Weggis and then a long hike up to the top, 1797 m above sea level. Most people walked up, though Queen Victoria was sedately carried in a sedan chair and Mark Twain famously took three days not to see the sunrise at all.

Rigi became such a popular destination that the Swiss chose it as the location for Europe's first mountain railway, using the then revolutionary cogwheel technology. By 1873, tourists could travel all the way from lakeshore at Vitznau to summit on the *Rigibahn*, prompting even more visitors and a hotel boom on the mountain.

This map dates from the early 1860s, so a few years before the railway opened. The four Rigi hotels then existing are shown in the corner vignettes, with the Kulm, or summit, Hotel in the top left. That one would grow ever larger and grander, almost obscuring the view, until it was demolished in 1952.

In the ring around the map are all the mountains visible from the summit, shown as extensions of the map itself. The Eiger, Mönch and Jungfrau peaks can be found in the bottom left.

About **Rudolf Gross** (1808-1880s?) very little is known, other than he was a German mapmaker who lived in Stuttgart then Basel. But he drew this wonderful map that was sold in Lucerne as a souvenir for visiting tourists.

Karte und Panorama vom **RIGI** Gezeichnet von Rudolf Gross.

Verlag v. C. Hunggüne-Schmid in Zürich.

41   The first inter-rail tourists

Travelling abroad was once something only rich people could afford. They went on their Grand Tours while the rest stayed at home and worked. Then along came Thomas Cook.

In June 1863 he took a group of British tourists from London to Lucerne and back. It was the first conducted tour of Switzerland, and it launched the era of mass tourism. Using the trains and group tickets, Cook made foreign travel accessible and affordable for the British middle classes, who flocked to the Alps in their thousands. This tourist influx helped finance new train lines, which encouraged more tourists to come; it was a perfect circle of development.

To make things easier for his clients Cook created special train tickets that were bought in advance in London. They were valid for a month and covered the journey out and back, and any chosen routes within Switzerland could be added on. In other words, an inter-rail pass but 150 years ago.

To help customers know which lines were covered by their tickets, Cook printed his own rail maps and timetables, which helpfully includes coach and steamship routes as well. This Swiss map appeared in the very first issue of *Cook's Continental Time Tables & Tourist's Handbook*, published in London in March 1873, and priced one shilling.

According to that timetable, it took 50 hours and 33 minutes to travel from London to Geneva, via Newhaven, Dieppe and Paris. The special fare for a return ticket, valid one month, was £7 and 5 shillings in 1st class or £5 and 8 shillings in 2nd (respectively about £550 or 750 francs and £400 or 550 francs today).

Thomas Cook (1808-92) was a Baptist preacher from the English Midlands. His first tour was a train trip for fellow anti-alcohol campaigners in 1841. After his Swiss success, his travel agency became famous for Nile cruises and world tours.

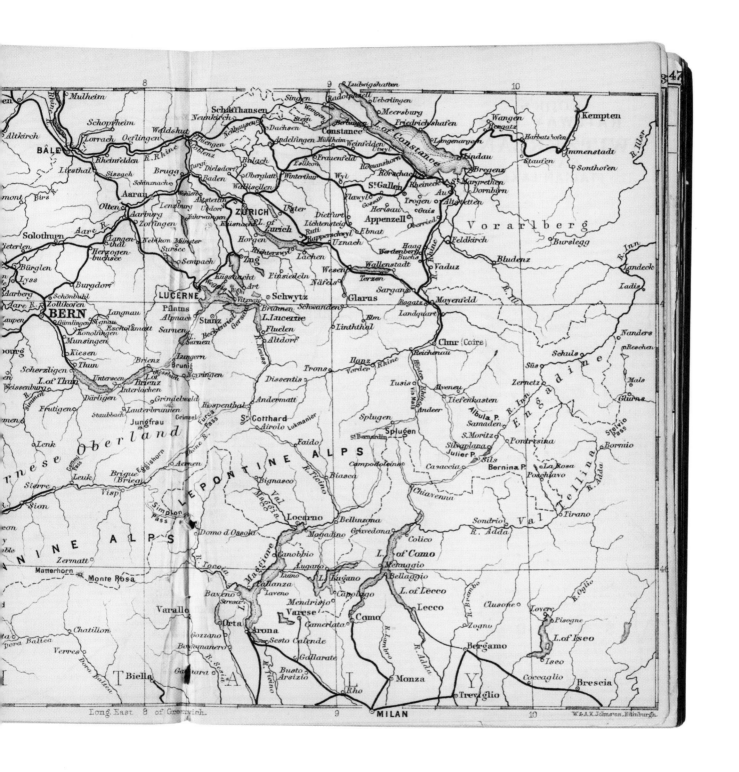

HUBACHER S·A· BERNE

42 + 43

Two posters for two tunnels

On 22 May 1882 the greatest Swiss engineering project of the 19th century opened with a grand fanfare.
The Gotthard Tunnel was in business! Switzerland finally had a rail network that had conquered the Alps.

▶

42
+
43

Two posters for two tunnels

◀

On 22 May 1882 the greatest Swiss engineering project of the 19th century opened with a grand fanfare. The Gotthard Tunnel was in business! Switzerland finally had a rail network that had conquered the Alps so that north and south were no longer separated by mountains or by snow.

At 1151 metres above sea level the 15km-long tunnel was the high point of the new Gotthard line from Lucerne to Chiasso. It had taken ten years to build, with the difficult terrain and bad conditions leading to many workers' deaths, including that of the project's architect Louis Favre. In 2016 the new Gotthard Base tunnel will open deep below the existing one (see page 158).

(see page 158)

With the Gotthard Tunnel open, the cantons in western Switzerland lost both trade and tourists, so responded with a tunnel of their own. Partly financed by the French government, the Lötschberg Tunnel connected Kandersteg in Canton Bern with Goppenstein in Canton Valais.

Operated by the Bern-Lötschberg-Simplon railway (or BLS), the 14.6km long tunnel opened on 15 July 1913 after 66 months work and 64 deaths. It was electrified from the beginning and slashed travel times to Italy, but was also superseded by a faster Base tunnel in 2007.

These two beautiful posters from the early 20th century show how maps have always been used to promote Swiss railways. Both clearly illustrate that their respective lines connect northern and southern Europe, though the lady looks like she has intestines made up of the Gotthard line. Sometimes a map really is worth a thousand words.

The Gotthard-Bahn poster, artist unknown, was printed in 1902 by Officine d'Arti Grafiche Chiattone in Milan, Italy.

The BLS poster, also artist unknown, dates from 1920 and was printed by Druckerei Hubacher in Bern.

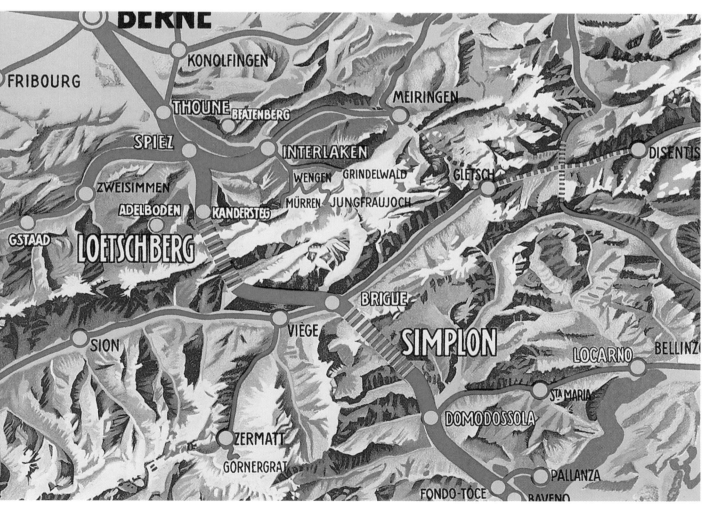

44   How the railways developed

The Swiss rail network got off to a bumpy start.
The first line, from Zurich to Baden (known as the
Spanisch-Brötli-Bahn), opened in 1847 but after
that the cantons could not agree on any lines or their
financing. So none was built.

It took a new federal government and the railway
law of 1852 to kick-start the construction of more lines,
but once they had started, there was no holding the
Swiss back. Railways sprang up everywhere, including
up and under the mountains.

The network's development until the First World War
is shown perfectly in this colour-coded map. From the
single black line of 1847 to the yellow-blue-red spaghetti
of the first big building craze that formed the basis
of today's network. Then the great engineering feats of
the *Rigibahn* (opened 1871-3) and the Gotthard Tunnel
(1882) under the Alps. Finally pushing the network to its
limits, be that up to Jungfraujoch, at 3454m high, or
over the Bernina Pass in Graubünden.

Initially Swiss railways were built by private companies
operating on cantonal concessions, but as more lines
were built, a national player was needed. In 1898 a
referendum approved the creation of the Swiss Federal
Railways (or SBB) and its first train pulled out of Bern
station on 1 January 1902.

This map of 1913 shows that by then most of today's
network had been built, with over 5,000km
already in service. The biggest missing link is the route
between Valais and Graubünden, the last part of
which opened in 1926. A few lines have also since disappeared, eg the orange ones to Bignasco
and Mesocco in Ticino.

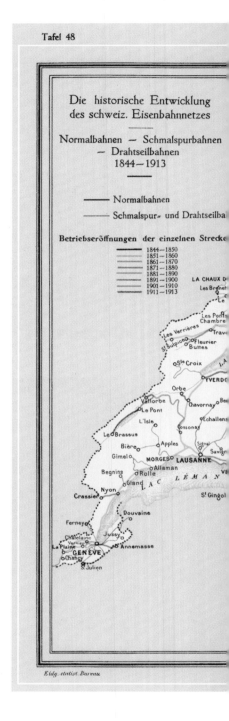

This map from the Swiss Federal Statistics Office was published in 1913, just as the
peak era of railway building came to an end. Then came the First World War, causing the
(temporary) collapse of tourism and further railway building projects.

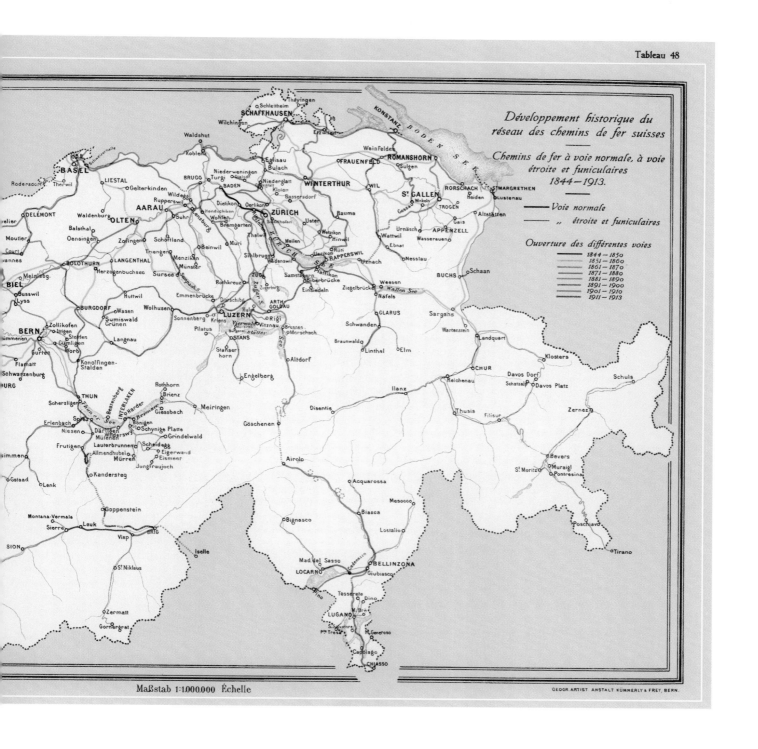

45　The Swiss rail network in 3D

This is not a computer-generated map of Switzerland's railways; this is a map from a 1915 transport atlas. Yes, 1915 when each of these lines had to be created by hand, which makes its depiction of the mountainous Swiss railways with a 3D-effect all the more impressive.

Although the Swiss were a little late building their railways (see page 134), they caught up for lost time quickly. Nearly every line on this map was built in the 60-year period from 1852 onward, including those going over and under the Alps, all represented wonderfully in this map.

In northern Switzerland it's hard to make out individual lines, thanks to the density of the network and the similar altitudes involved. But in the Alps, the peaks are plain to see. Almost in the middle is the vertiginous spike of Jungfraujoch, Europe's highest train station (see page 154), while on the far right are the heights of the Bernina Line, the highest rail crossing in Europe that doesn't use a rack railway (see page 146).

At the bottom of the map in Valais is the steep line to Gornergrat above Zermatt, which opened in 1898 as Switzerland's first fully-electrified cog railway; it peaks at 3089 metres up. To its left on the other side of the Rhone valley is the railway up to Leukerbad, which was brand new in 1915 but closed 52 years later. Also now gone are the lines to Bignasco and Mesocco, shown here as the outer arms of the "trident" in Ticino.

The total length of network shown here is 5,064km, or about 300km less than today. Of that, roughly 30% was in narrow gauge and rack railways, while tunnels made up 162km.

Horizont = Meereshöhe

The map is from the *Graphical-Statistical Transport Atlas of Switzerland* published by the Swiss Post and Railways Department in 1915. It has two scales – 1:250,000 for horizontal distances and 1:10,000 for vertical – with the horizon drawn at sea level.

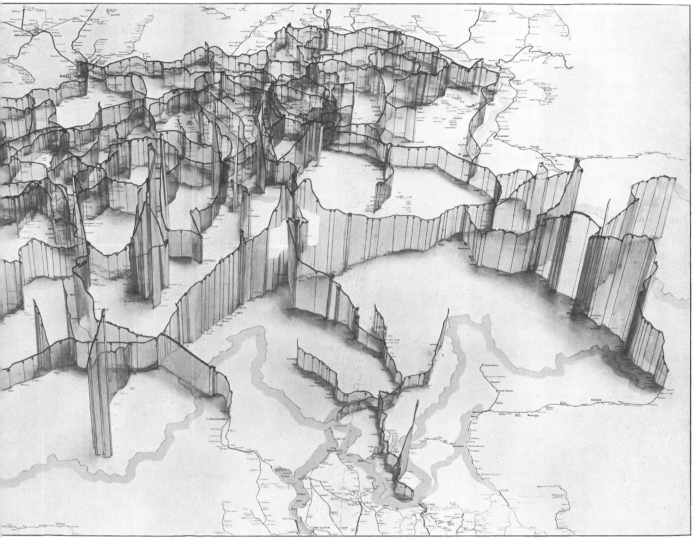

gen = 1 : 250 000, Höhen = 1 : 10 000   Carte originale : longueurs = 1 : 250 000, hauteurs = 1 : 10 000   Horizon = Niveau de la mer

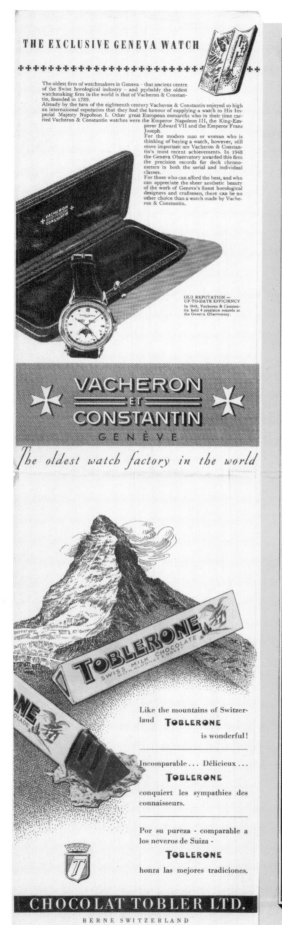

SWISSAIR ROUTES

VOLS SPECIAUX
SPECIAL-FLIGHTS

Projection van der Grinten
L'échelle n'est exacte que le long de l'équateur
1:32 000 000

46

Flying with Swissair in 1949

Geneva to New York in 23 hours, including stops in Shannon and Gander, Newfoundland: that was the new scheduled service from Swissair in the spring of 1949.

▶

46   Flying with Swissair in 1949

◄

Geneva to New York in 23 hours, including stops in Shannon and Gander, Newfoundland: that was the new scheduled service from Swissair in the spring of 1949. It was the start of long-haul flights for the Swiss national airline, and proved so successful that Rio followed in 1954, via Lisbon, Dakar and Recife.

But, as this route map shows, most Swissair services in 1949 were still European, though there were flights to Cairo and Lydda (the old name of Tel Aviv's airport). One notable absence is any flights to Germany, but in 1950 Stuttgart, Hamburg and Munich were back on the schedule; Berlin had to wait a while longer.

The direct flight between Zurich and London took almost 2.5 hours and had a particular significance. In July 1945 it had been the first postwar flight from Switzerland, specially arranged to bring the English football team over for a match. There had been no other way of getting them to the game. Switzerland won 3 - 1, and then flew the losers home.

Swissair was founded in 1931 with ten planes (and 86 seats) and 64 staff, including ten pilots. In 1934 it was the first European airline to hire a flight attendant, a lady called Nelly Diener who sadly died the same year in a plane crash. By 2001 Swissair flew to 218 cities, carried 19.2 million passengers and employed 71,900 staff.

But Swissair was in financial difficulty after overstretching itself and the after-effects of the September 11 attacks affected it more than most other airlines. On 2 October 2001 its planes were grounded completely, and on 31 March 2002 it ceased to exist.

Swissair network map from May 1949, published by Kümmerly & Frey, and including the new scheduled service to New York La Guardia. The flights and times might have changed but the adverts have not. Watches and chocolate: the same now as it was then.

47 Planning the new motorways

Switzerland's first stretch of motorway opened in June 1955, built as a by-pass between Lucerne and Horw (it only later became part of the A2). At that time no national plan for a motorway network had been agreed upon, even though Germany and Italy had been building their famous *Autobahn* and *autostradas* for over 20 years. It was time for action.

Three years later the Swiss government published its Strategy for the Swiss National Road Network, a six-volume report outlining its vision for the future of Swiss main roads. By 1980 there would be three classes of national roads, as this map shows: 571km of Class I (ie the main motorways), 559km of Class II (primary links) and 542km of Class III (Alpine roads & secondary links).

Upgrading and new construction was planned in two stages: first in 1960-9 and costing 2.9 billion Swiss francs, then 1970-9 costing 888 million Swiss francs. The lion's share of the money was for the new motorways, some of which were planned through the heart of Swiss cities (see overleaf).

The plan was approved by a referendum and work started soon after, with the first section of actual motorway opening in May 1962: the A1 at Grauholz in Canton Bern. The first motorway service station followed in 1967 at Kölliken in Aargau. By 1980 876km of motorway had been built; today the total is 1419km.

Not everything went to plan. There were several options for linking Cantons Bern and Valais, with the Rawil Pass between Lenk and Crans-Montana chosen over the Gemmi (see page 202). The Rawil route, with its 4km-long tunnel under the pass, is on the map but was never built.

Despite that, and the inner city motorways which also didn't materialise (see overleaf), this map is remarkably similar to a modern Swiss road map, although the 2015 version has more motorways, of course.

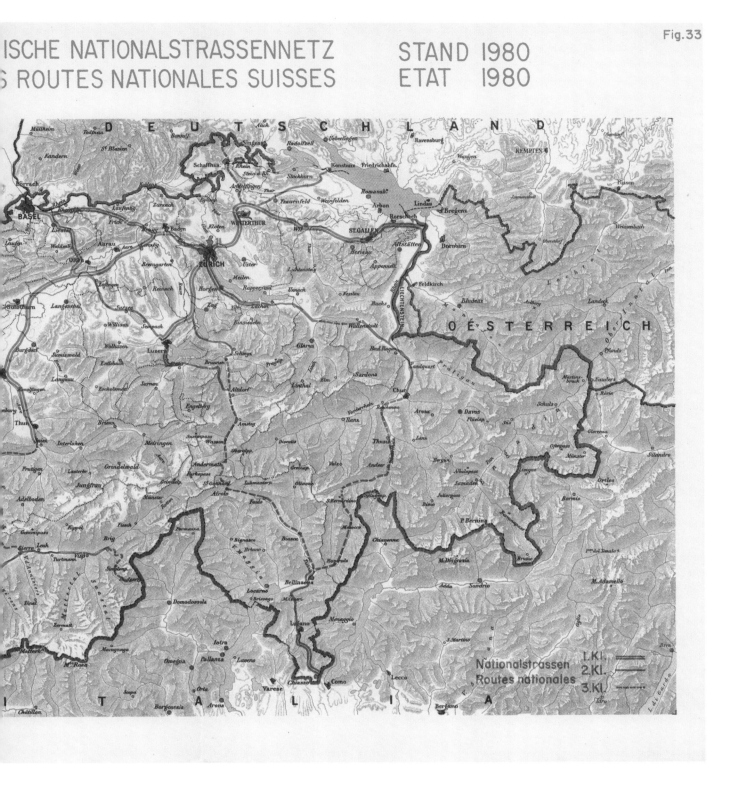

48
+
49

◄

Even back in 1955 the busiest spots were near the two biggest cities, with the main roads around Zurich notching up 8,707 vehicles per day, while Geneva wasn't far behind, with 8,003 per day. Both were streets ahead of Basel, on a mere 5,500. The report forecast that these figures would treble by 1980, taking Zurich's busiest road - the new A1 - to 25,800 vehicles a day.

To help manage traffic flows into city centres and alleviate congestion, Urban Expressways were to be built through all big Swiss cities, though they weren't designed with aesthetics in mind. In Zurich three Expressways would meet in a Y junction near the main train station, and then run through the heart of the city along the Limmat and Sihl rivers. But the project faltered against stiff local opposition and was abandoned in the 1970s; Zurich had to wait until 2009 for the alternative, a motorway round the city, to be completed.

The Geneva plans were less concrete, as the dotted-line variations on this map show. One option was for an Expressway right along the lakeshore past all the posh hotels; a very Fifties concept of building dual carriageways as a sign of modernity and progress. It didn't happen and the other option past the airport was built.

On a national level the project mostly went to plan while vehicle numbers rocketed from 252,000 in 1950 to 2.7 million in 1980 – and 5.7 million today. As for those traffic flow forecasts, they were surpassed by a huge degree: 142,200 vehicles per day now travel on Switzerland's busiest road, the A1 past Zurich.

The maps come from the 1958 government report on Swiss roads, and show how the main road network would look in 1980. After 20 years of construction there would be 1,672km of national motorways and main roads across the three categories.

Expressstrassen und Nationalstrassen
Routes expresses et routes nationales

Spätere Entlastungsstrassen: Zeit der
Ausführung noch unbestimmt
Routes de décharge ultérieures: date d'exécution
encore indéterminée

GENEVE
AUTOBAHNVERBINDUNGEN
LIAISONS DES AUTOROUTES

1 : 25 000

Expressstrassen und Nationalstrassen
Routes expresses et routes nationales

Routen im Studium
Routes à l'étude

50   A railway through the Alps

This is a train that runs through the mountains but isn't a mountain train. This is a railway that winds up to the highest rail crossing in Europe without cogs. This is a line that has been declared a Unesco world heritage site. This is the Bernina Line.

From glitzy St. Moritz up over the Bernina Pass and down into Italy, the Bernina Line is one of the most spectacular train journeys in Switzerland. It opened completely on 5 July 1910, six years after the equally scenic Albula Line, which links St. Moritz with Chur to the north. Together the two lines are Unesco-protected and form the route of the modern Bernina Express.

The railway climbs up from 1778m above sea level past snow-capped peaks and creaking glaciers to its highest point at Ospizio Bernina, 2253m up and the watershed between the Danube and Po basins. Then it's downhill all the way via Italian-speaking Poschiavo and over the border to Tirano in Italy, at a lowly 429m. There are no rack sections or cogwheels used, but instead tight s-bends and the 360° spiral viaduct at Brusio cope with gradients of up to 7%.

Despite heavy snows that are frequent at these altitudes in winter, the Bernina Line started year-round services soon after opening. But the line went bankrupt and in 1943 it became part of the Rhaetian Railway.

This playful map, which was used to publicise the line in the 1960s, is oriented with west at the top so that the whole 60.69km of the Bernina Line fits neatly on a page in landscape format.

Otto M. Müller (1913-2002) drew this lovely map but its exact date is unknown. One text says that the style of the bikini (on the lady beside Lake Poschiavo) suggests that it was published somewhere between 1950 and 1960.

Maloja

OBERENGADIN

Sils/Segl

Silvaplauna

P. NAIR   3057 m

P. ROSEG 3937 m

Corviglia

P. ZUPÒ 3995 m   P. BERNINA 4049 m

2486 m

ST. MORITZ

2286 m   Marguns

V. ROSEG

1778 m

Celerina

P. PALÜ 3905 m

Punt Muragl

Samedan

PALÜ-GLETSCHER

MORTERATSCH-GL.

1740 m

PONTRESINA

Surovas

1777 m

Diavolezza

2973 m.

Muottas
Muragl

Morteratsch

1899 m

MONTEBELLO

2453 m

P. LANGUARD 3261 m

ADRIA   SCHWARZMEER

ALP LANGUARD

A O

2091 m

A. Grüm

Bernina-Diavolezza

LAGO BIANCO

2085 m

P. ALBRIS 3166 m

POSCHIAVO

1014 m

1693 m

Ospizio 2257 m

Cavaglia

P. d. SENA 3075 m

P. TREVISINA 2823 m   L. D. SAOSEO

L. VIOLA

Cma. d. SAOSEO 3265 m

CHUR · ZÜRICH · BASEL

DAVOS

SCUOL/SCHULS-TARASP

W

S

N

O

O.M. MÜLLER

51 The Zurich map for men

It's certainly not a normal tourist map. Instead of the main sights, there are strip clubs and gay saunas; where to eat is replaced by where to meet for adult fun. This is a City Map for Men, with everything in it tested and reviewed.

Red dots signify 'leichte Gewerbe', literally "light trade" where "accommodating ladies are waiting". That means erotic salons and saunas (ie brothels in everything but name), but also streets and districts where prostitutes can be found.

Yellow dots are for "people from the other platform", which is made clearer by the 'Homos' icon of two men hugging. Places include Barfüsser in Niederdorf, once the oldest gay bar in Switzerland. Opened in 1956, it is now a sushi restaurant. Blue dots offer "spicy evenings and frivolous nights", from strip shows and cabaret bars to discos and nightclubs. All of these are for customers who are "an adventurous Casanova".

The map's highlight is the range of icons for services and clientele. So the girl with perky red hair and long legs shows streets where the prostitutes are "young", whereas the blousy hooker with a blue rinse and sagging breasts signifies a "rather worn-out" strip. The Go-Go stripper looks distinctly unhappy with her lot and the transvestite was clearly the inspiration for Conchita Wurst.

The map has a remarkably liberal and open attitude to sex tourism, especially given the era. It even gives price guides for services, from 50 francs (or £15 then) for "scarcely ten minutes' pleasure" to "from 300 francs for special requests in S&M".

This Zurich map was one of 17 City Maps for Men produced by Monika Dülk Verlag in the 1970s and 1980s. Most of the cities were German, eg Berlin or Munich, but there were also Amsterdam, Paris and Bangkok maps. This one of Zurich was the 5th "improved" edition.

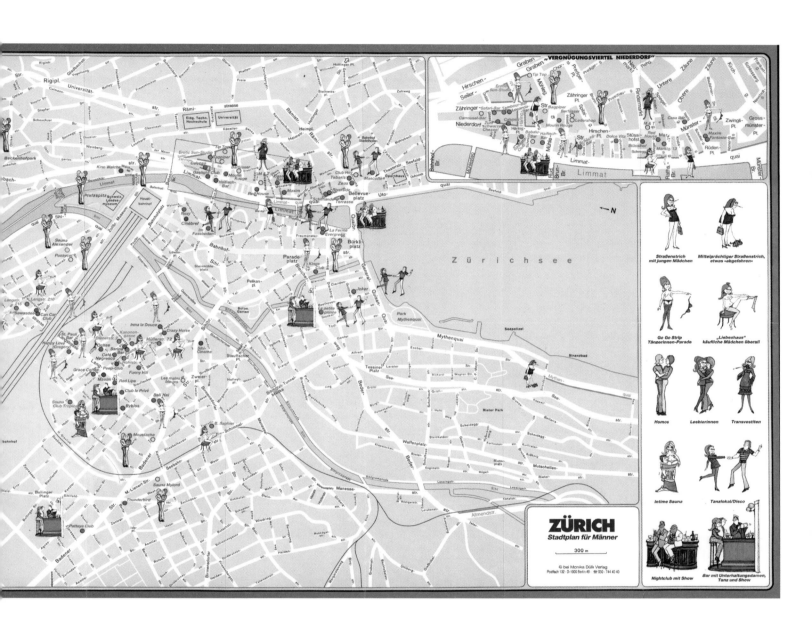

Plan du réseau jusqu'au 1er juin 1991

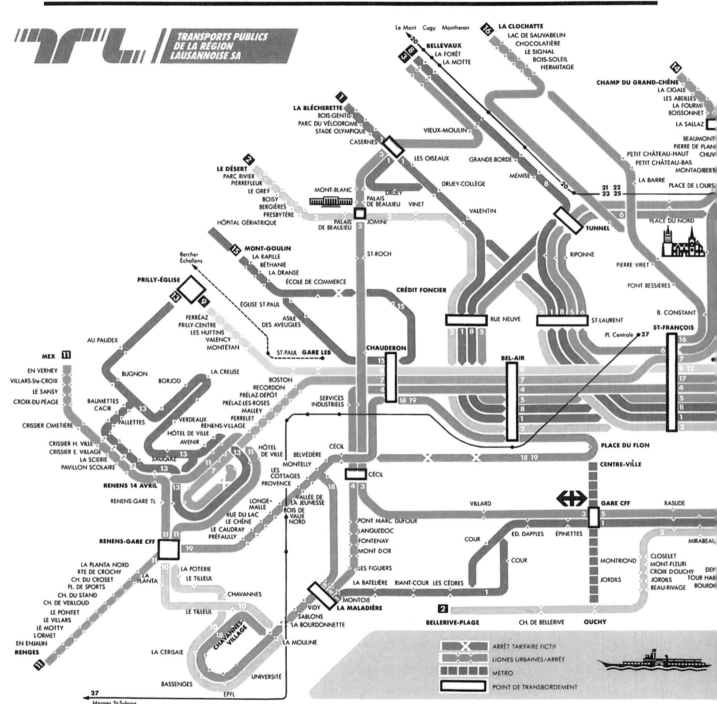

52   A singular metro line

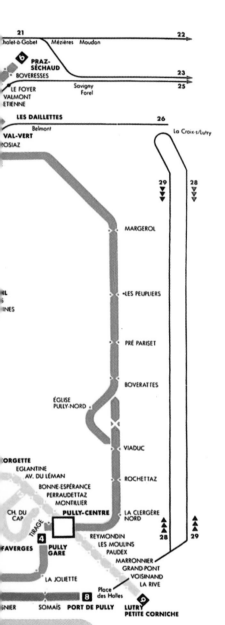

Lausanne is Switzerland's fourth largest city but is the only one with a metro, one that has its origins in 1877. Today there are two lines, M1 and M2, but the system began with Switzerland's first public funicular, the Lausanne-Ouchy line.

On 15 March 1877 a funicular railway opened between Ouchy down beside Lake Geneva and Flon in Lausanne city centre, with a stop at the main train station (Gare CFF on this map). Known as *La Ficelle*, or The String, it covered an altitude difference of 106 metres over a distance of 1,482 m, with a maximum gradient of 11.6%. It was busiest on Sunday afternoons when city folk went down to promenade along the lakeshore. A second funicular was added in 1879 but only along the short distance between the CFF station and Flon to relieve overcrowding.

In 1958 *La Ficelle* was replaced with a rack railway and rechristened the *Métro*, though many locals still used the old name. It was now a single track railway, with a crossover point at Montriond, and on this map is the purple dotted line linking Ouchy and Flon.

In 1991 the Ouchy metro was managed along with a then-new light-rail line, known as M1, but it finally closed in 2006 to make way for Switzerland's first totally automatic metro. Most of the old stations vanished and the line was extended out to Les Croisettes in Épalinges to the north. This new M2 underground line opened on 18 September 2008, with a total length of 5.9km and an impressive height difference of 338m between Ouchy and Épalinges.

Official Lausanne transport map valid until 1 June 1991, the day before the new M1 line opened to the public. M1 isn't actually an underground but a 7.8km-long light rail line running from Renens in the west to Flon in the centre, where it connects to the M2.

53 A 700th birthday present

In 1991 Switzerland was officially 700 years old so the Swiss created
a new hiking path to celebrate. It might not sound like the most
exciting celebration but it is typically Swiss as it is modest, involves the
national pastime of hiking, and needs attention to detail. And it was
cheaper than Plan A, a national Expo fair, which had been rejected by
the voters.

The path was a very precise undertaking, planned down to the last
millimetre, literally. Every 5mm of path represents one inhabitant of
Switzerland at that time, meaning that the population of (almost)
seven million was transformed into a 35km hiking trail that winds
around Lake Uri in central Switzerland

Each of the 26 Swiss cantons has its own section of the path, based
on the cantonal populations in 1991. That means that Zurich, the most
populous, bags the longest stretch while the smaller ones, such as
Glarus, manage only a few hundred metres. Stone markers along the way
mark the boundaries between the cantonal sections, shown on this
map by the coats of arms.

The path's starting point is Rütli meadow, considered to be Switzerland's
birthplace. According to the legend, it was here in 1291 that men from
the first three cantons swore an oath of allegiance against the Austrians,
an event that is commemorated every year on Swiss National Day,
1 August.

From Rütli the cantons are set out along the path in the order that they
joined the Confederation, starting with those first three – Uri,
Schwyz and Unterwalden (separated into Nidwalden and Obwalden) –
and ending with Jura, which became the last canton in 1979.

The Swiss Path was inaugurated in 1991 and runs from Rütli in Canton Uri to Brunnen
in Canton Schwyz. It is 35km in length, representing both 700 years of Swiss history and
seven million inhabitants.

Gr. Windgällen
3187

Bristen
3072

Uri-Rotstock
2928

Gitschen
2511

San Gottardo

Amsteg

Erstfeld

Brüsti

Haldi

Eggbergen

Attinghausen

Schattdorf

Bürglen

Seedorf

St. Lazarus

A Pro

St. Ulrich

Altdorf

Unterdorf

Reussdelta

Oberbauenstock
2117

Reuss

Bolzbach

Isenthal

Flüelen

Rodenz

Scheidegg

Usserdorf

Gruonbach

Isleten

Niederbauen-Kulm
1923

Axenflue

Adams Rüti

Tellskapelle

Bauen

Zwyssighaus

Tellsplatte

St. Idda

Grawegg

Wissig Hostet

Herbizigegg

U r n e r s e e

Beroldingen

Frohwald

Seeli

Sisikon

Ried

Schiferenegg

Oberdorf

Emmetten/Beckenried

Tannen

Rütli

Kulm-Sonnenberg

Zingel

Maria-Sonnenberg

St. Franz-Xaver

Rütliwald

St. Gallus

Chilendorf

St. Michael

Axenstrasse

Seelisberg

Morschach

TSB Station

Axenstein

Wolfsprung

Chänzeli

Schillerstein

Brüebi

Brunnen

Treib

Gersau / Beckenried / Luzern

Bundeskapelle

Muota

V i e r w a l d s t ä t t e r s e e

Ingenbohl

PLATZ DER AUSLANDSCHWEIZER
PLACE DES SUISSES DE L'ETRANGER
PIAZZA DEGLI SVIZZERI ALL'ESTERO
PLAZZA DALS SVIZZERS A L'ESTER
PLACE OF THE SWISS ABROAD

54 + 55   Up to Europe's highest train station

Same view, different century. These two postcards both use relief maps to highlight the magnificent railways of the Bernese Oberland, which were mainly built to take British tourists ever higher up the mountains. More tourists meant more money for more trains so why not let those tourists do the marketing for you by sending a picture-postcard home?

The older view is from about 1910 with rack railways snaking into the hills behind Interlaken. The train had reached Grindelwald and Wengen in the 1890s, at the height of the Belle Epoque tourist boom in Switzerland, and then it was ever upwards. A railway was planned all the way to the summit of Jungfrau; here it is shown as having been started above Kleine Scheidegg but the end of the line is dotted. It never reached its goal but stopped at Jungfraujoch, Europe's highest train station (3454 metres up), which opened on 1 August 1912.

The world's first aerial cable car can be seen on the left, labelled 'Wetterhorn Aufzug'. Opened in 1908, it was originally planned to reach the top of the Wetterhorn but the First World War halted the project with only one stage complete, and even that was dismantled in the 1930s; the disused upper station can still be seen part-way up the mountain.

Despite that failure, cable cars went on to conquer many mountains: in the modern view they are shown by black lines swinging up to the summits. Those Victorian mountain train lines are all still there, now shown in red, and not many others have been built. And Jungfraujoch is still Europe's highest train station.

Two postcards from the archives of the Jungfrau Railway. The line was the brainchild of Adolf Guyer-Zeller and it took 16 years to build the 9.34km from Kleine Scheidegg up to Jungfraujoch. Guyer-Zeller died in 1899, 13 years before the line opened.

56 Mountains of art

With 48 named peaks over 4000 metres, Switzerland can justifiably be called the roof of Europe. Over the centuries its mountains have inspired as many artists as adventurers but it's not often that the names of the mountains become the work of art.

This typographic topographic map of Switzerland by Ursula Hitz uses the names of 41 Swiss peaks to form the outline of the country. They are shaped and knitted together like a free-form jigsaw puzzle, but with a focus: the position of each peak is marked by the white triangle near its name, along with its altitude.

Hitz's first typographic map was of London, with the names of boroughs creating a complete city map. In Switzerland she used the topography of the country instead, starting with the highest point in each canton, which produced a base of 25 names – Appenzell Innerrhoden and Ausserrhoden share Säntis (2503m). The lowest cantonal highest point is Les Arales (516m) in Geneva, while the highest is also the highest point in Switzerland at 4634m: Dufourspitze in Valais (see page 46).

Then she added some famous peaks that simply couldn't be left out of such a Swiss map, for example Eiger, Mönch, Jungfrau, Matterhorn, Pilatus and even Gurten (only 858m) near Bern. Finally, one mountain chain in eastern Switzerland, Churfirsten, that was a memorable part of her childhood.

Each name was created individually and then shaped to fit together as a carpet of words. Like all of Hitz's maps, it was created purely in digital form but was then printed for sale.

Ursula Hitz (born 1980) grew up on a dairy farm in Canton St. Gallen and then studied graphic design. Over a decade ago she moved to London to improve her English and gain experience as a graphic designer.

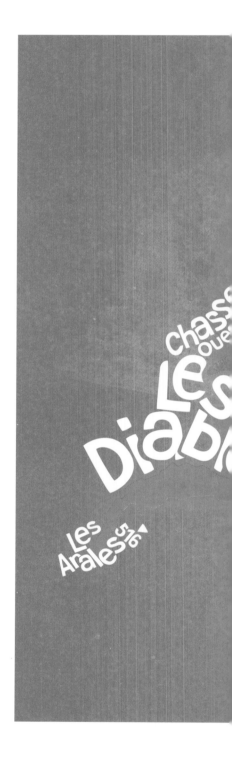

©URSULA HITZ

57   The world's longest rail tunnel

Grinding its way at a rate of 40 m a day, the giant boring machine took almost eleven years to carve through the rock beneath the Gotthard Massif. Finally at precisely 14:17 on 15 October 2010 the breakthrough was made and the world's longest rail tunnel was a step closer to completion.

The first Gotthard tunnel was a wonder of technology when it opened in 1882, but the winding scenic route up to the tunnel at 1 151 m above sea level was no longer compatible with the modern age of fast trains. So a new tunnel was planned to flatten out the trip under the Alps.

At 57.104km, the new Gotthard Base Tunnel will be longest rail tunnel in the world when it opens in December 2016; it will also be the world's deepest, with up to 2 300m of rock above it. Trains will travel at a maximum speed of 250 km/h through twin tunnels built 40m apart, reaching a highest altitude of only 549m.

A second tunnel, the 15.4km-long Ceneri Tunnel, will iron out the wrinkles down to Lugano and is scheduled for service in December 2019. Along with the Lötschberg Base Tunnel to the west (34.6km, opened in December 2007), the tunnels are part of the ambitious New Rail Link under the Alps (NRLA) project – with a total price tag of 18.5 billion Swiss francs.

Building the new Gotthard Tunnel is a massive undertaking, employing 2 600 people for 4 million man hours of work. Over 28 million tonnes of rock has been excavated and replaced with 4 million m³ of concrete and 1.4 million tonnes of cement to create the new twin tunnels.

Three aspects of the NRLA from an AlpTransit brochure: the route of the new tunnel compared to the old line; a geological cross-section of the Gotthard Massif; and a profile with altitudes, with the existing 1882 tunnel as a short dotted line in black between Göschenen and Airolo.

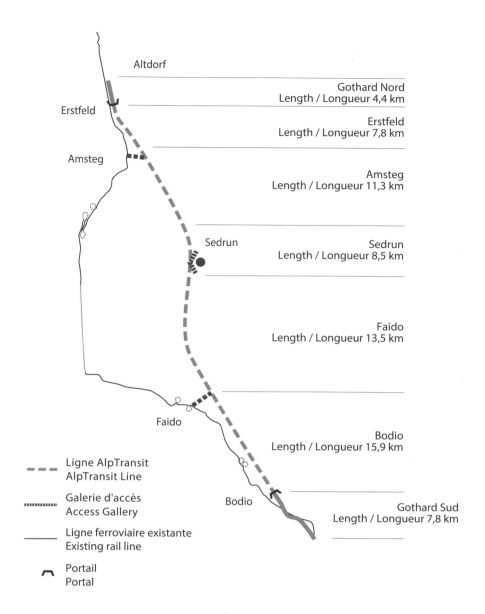

Altdorf

Erstfeld

Amsteg

Gothard Nord
Length / Longueur 4,4 km

Erstfeld
Length / Longueur 7,8 km

Amsteg
Length / Longueur 11,3 km

Sedrun

Sedrun
Length / Longueur 8,5 km

Faido
Length / Longueur 13,5 km

Faido

Bodio
Length / Longueur 15,9 km

Bodio

Gothard Sud
Length / Longueur 7,8 km

- - - Ligne AlpTransit
AlpTransit Line

······· Galerie d'accès
Access Gallery

—— Ligne ferroviaire existante
Existing rail line

⌐ Portail
Portal

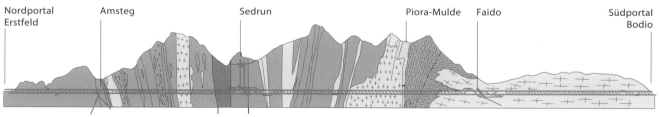

Nordportal
Erstfeld

Amsteg

Sedrun

Piora-Mulde

Faido

Südportal
Bodio

The Tavetscher Intermediary Massif
Massif intermédiaire de Tavetscher

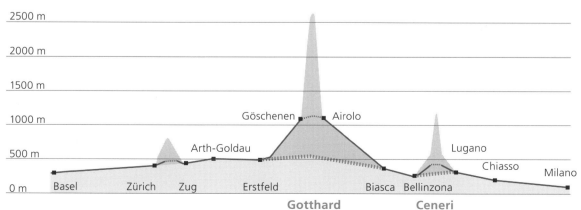

2500 m

2000 m

1500 m

1000 m

500 m

0 m

Göschenen   Airolo

Arth-Goldau

Lugano

Chiasso

Milano

Basel   Zürich   Zug   Erstfeld   Biasca   Bellinzona

Gotthard   **Ceneri**

People
&
Power

Explanation and
information, from livestock
and language to population
ups and downs

58   The heart of Swiss democracy

With the new constitution of 1848 Switzerland became a federal state and needed a capital city. In November Bern was chosen, though it was diplomatically named the Federal City rather than a capital.

A first *Bundesrathaus*, or federal town hall, opened in central Bern in 1857, followed by a second building in 1892; today they are known as the West and East Wings, respectively. It wasn't until 1894 that construction began on the central *Bundeshaus*, or Federal Parliament Building, standing between these two wings and crowning the whole ensemble with a dome. This is the plan of that central parliament.

The architect, Hans Wilhelm Auer, saw this building as the embodiment of the new Switzerland, using 38 Swiss artists, 173 Swiss firms and stone from 13 cantons to create his masterpiece. Every statue and painting shows something from Swiss history, such as the cantonal coats of arms under the dome and above the central hall that is shaped like a Swiss cross with staircases rising up to the two houses of parliament.

At the front of the building (ie at the bottom of this plan) is the Council of States, the smaller of the two chambers with 46 seats, two for each canton. Across the central hall is the larger National Council, with 200 seats allocated according to the cantons' population. This chamber's wide arc is shadowed by the *Wandelhalle*, or lobby, stretching 44 m along the curved back of the building (ie at the top of the plan) with views out to the Alps.

The *Bundeshaus* opened on 1 April 1902, having cost 7.2 million francs to build, or 700 million (£450 million) in today's money.

Hans Wilhelm Auer (1847-1906) was from Canton St. Gallen. He first built the second wing of the *Bundesrathaus* and in 1891 was commissioned to construct the new domed *Bundeshaus*; these are his plans from 1901.

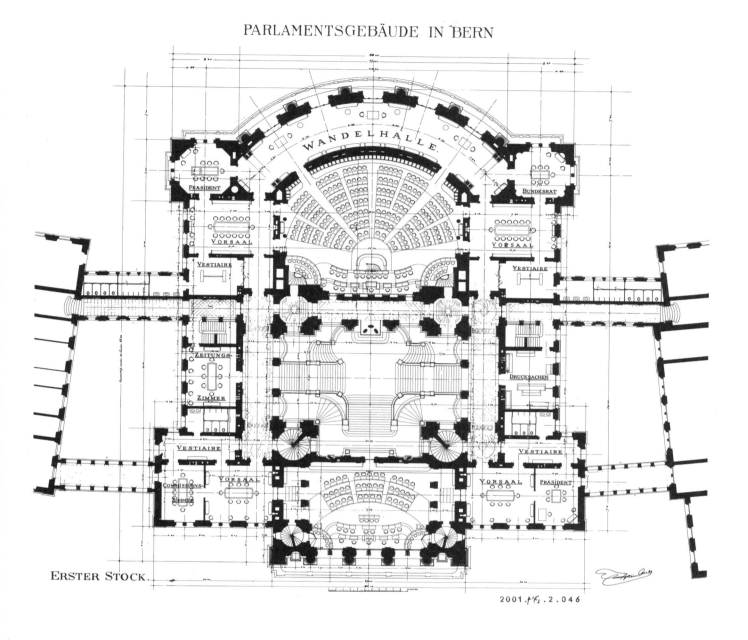

PARLAMENTSGEBÄUDE IN BERN

ERSTER STOCK.

2001..2.046

59   The shadows of death

Even without knowing the subject matter, it's clear straightaway that Romandie and northern Switzerland are more affected than the rest of the country. Once you learn that this map shows the suicide rate in Switzerland at the beginning of the 20th century, you immediately wonder about those areas in particular.

Sadly there is no clear answer as to why some districts (those with the dark shades) had rates twice the national average, and others none at all. Suicide figures were generally lower in rural (paler) areas than urban (darker) ones, but that may also be down to less accurate recording, for example noting the death as an accident rather than suicide. That could also help explain the lower figures in Catholic cantons, where accident rates were higher than normal.

The map covers the years 1901-10, when the number of suicides hovered around 800 a year, with a peak of 849 in 1904. The accompanying notes detailed the methods used: men were more likely to hang themselves (45%) or shoot themselves (25%), whereas for women almost exactly the same percentages were for drowning and hanging. Poison was used more often by women (10%) than men (3%).

Today the total number of suicides is higher (1,037) but at a much lower rate with regard to the population: 1.35 per 10,000 people, compared to 2.2 as shown on this map. Hanging has become the most common method for both sexes (31% overall), but poison is now used more often than 100 years ago: 19% of women and 11% of men. The urban-rural divide is still the same, but the differences between the language areas have evened out.

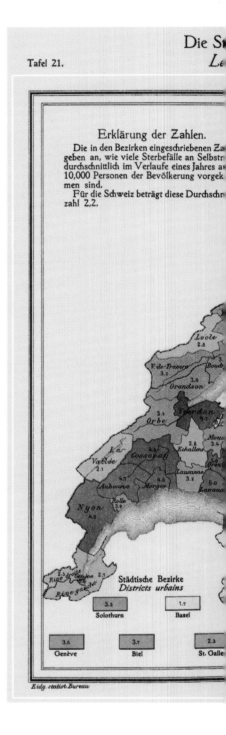

The map first appeared the *Graphical-Statistical Atlas* (Federal Statistics Office, 1914) and shows suicides per 10,000 people in each district. Lavaux, Oron and Cossonay (all in Vaud) had the highest rates but Gersau, Hérens and 'Münsterthal' had none at all.

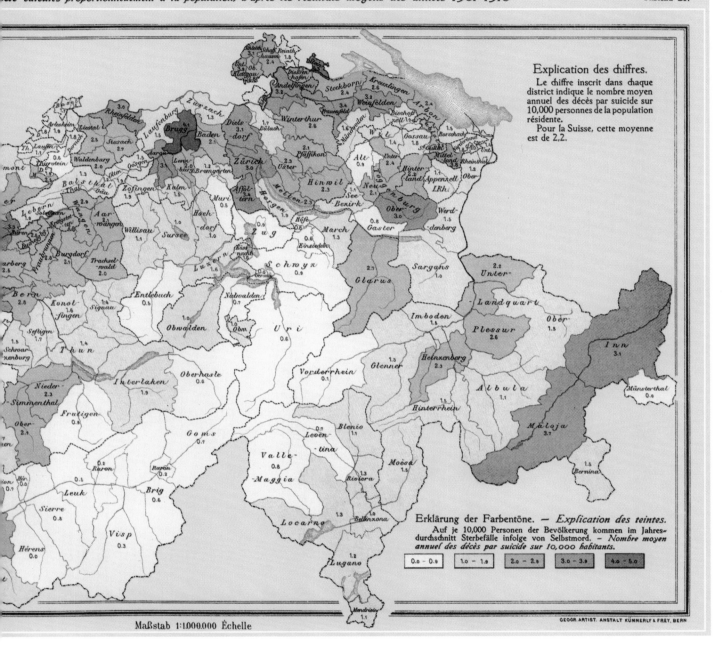

von Selbstmord im Verhältnis zur Bevölkerung, nach den Durchschnittsergebnissen der Jahre 1901–1910

ide calculés proportionnellement à la population, d'après les résultats moyens des années 1901–1910

Tableau 21.

Explication des chiffres.

Le chiffre inscrit dans chaque district indique le nombre moyen annuel des décès par suicide sur 10,000 personnes de la population résidente.

Pour la Suisse, cette moyenne est de 2,2.

Erklärung der Farbentöne. — *Explication des teintes.*

Auf je 10,000 Personen der Bevölkerung kommen im Jahres-durchschnitt Sterbefälle infolge von Selbstmord. — *Nombre moyen annuel des décès par suicide sur 10,000 habitants.*

| 0.0 – 0.9 | 1.0 – 1.9 | 2.0 – 2.9 | 3.0 – 3.9 | 4.0 – 5.0 |

Maßstab 1:1.000.000 Échelle

GEOGR. ARTIST. ANSTALT KÜMMERLY & FREY, BERN

60   Something old, something new

In the 20th century getting married became less popular, as these two maps show. They both use the same basic data, the number of marriages in Swiss districts, but the old map is from 1901-10 and the new exactly a century later. The national average of 7.5 marriages per 1,000 people in the first map fell to 5.4 by 2001-10.

The earlier map reveals a starker urban-rural divide, with most dark areas (ie with higher marriage rates) in the cities. In fact, all three districts with the highest rates were urban: Geneva on 10.7 marriages per 1,000 people followed by Tablat in Canton St Gallen (10.4) and St Gallen itself (10.2). Rural districts were generally paler, especially in mountain areas; in Valais two failed to break the 5.0 barrier - Goms and Raron.

By 2001-10 urban rates were lower and the city-countryside gap had narrowed noticeably. Only two districts (Zurich and Höfe in Canton Schwyz) scored over 7, while nearly half the districts had a rate under 5; only four were white, with under 4 marriages per 1,000 people.

Behind the maps' raw data is also the changing nature of marriage. In 1901-10 the average age for getting married for the first time was 28.6 for men and 26.3 for women; a century later that had risen to 31 for men, 28.7 for women.

More dramatic was the drop in marriages between two Swiss people, perhaps only natural given the rise in foreigners living in Switzerland. Whereas 78% of unions were with a Swiss bride and a Swiss groom, in 2001-10 it was only 51%. As for divorces, they have boomed, soaring from 0.4 per 1,000 people to 2.2.

The first map is from the *Graphical-Statistical Atlas* (Federal Statistics Office, 1914) using data from 1901-10. The second is from the 100th anniversary edition of the Atlas, using the same districts and shading for data from 2001-10.

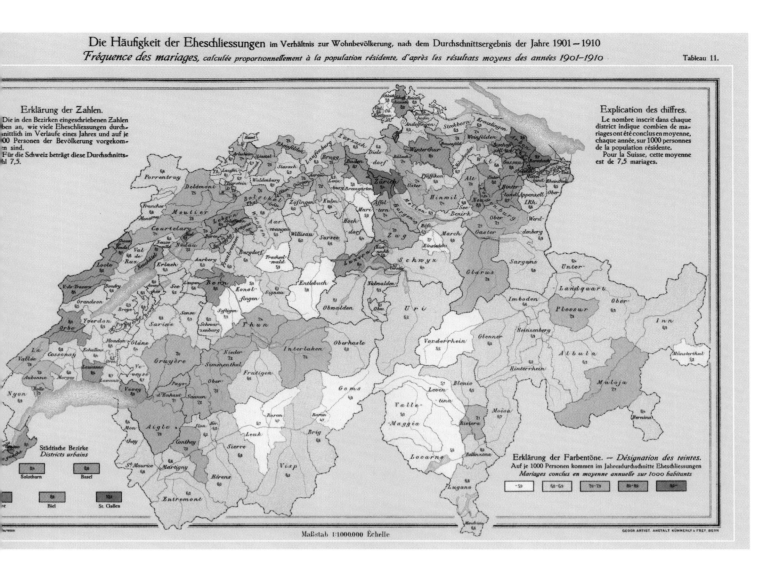

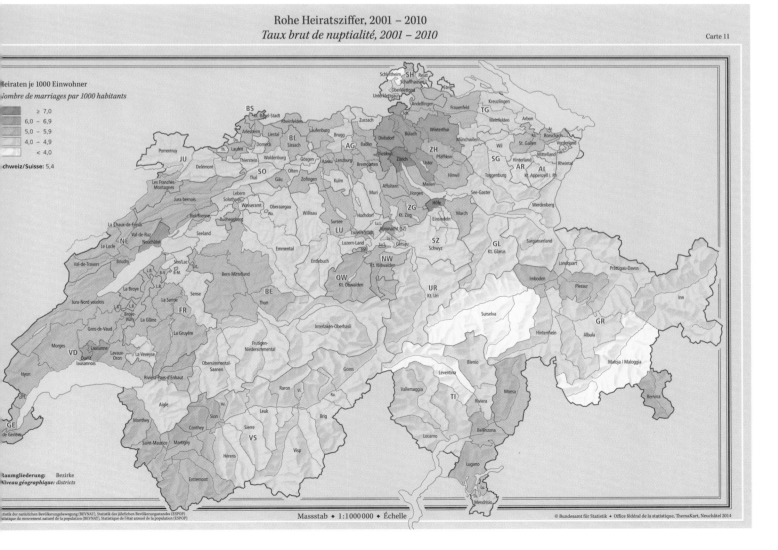

61   Comparing births across a century

Two maps with the same subject – the birth rate in each district of Switzerland – and the same scale – the number of live births per 1,000 people – but set 100 years apart. And what a difference! All the dark shades (ie districts with high rates) in the first, older, map have vanished in the second and the whole country is generally paler.

The birth rate across Switzerland collapsed from an average of 26.9 births per 1,000 people in 1901-10 (shown in the first map) to only 9.9 exactly a century later. People simply stopped having so many babies, partly because infant mortality rates fell. Some of the most spectacular drops were in districts with the highest birth rates, ie the darkest areas.

In 1901-10 the three highest-rated districts were Sense and Veveyse in Canton Fribourg (38 and 37 respectively per 1,000 people), and Riviera in Ticino (37.6). By 2001-10 those three districts' rates were drastically lower: Riviera with 9.6, Sense with 9.7 and Veveyse with 11.5.

A smaller change was a levelling out of the differences between Catholic and Protestant cantons. Catholic areas still usually have higher rates than Protestant ones but the disparity is less marked. And the fall was less dramatic in some Protestant areas.

Protestant Vaud and Geneva had the lowest birth rates in the country in the first map: 17 per 1,000 in Geneva and low 20s across much of Vaud; 100 years later they had some of the highest, because their rates had dropped by less. So Ouest lausannois was top with 12.3, Nyon third on 11.7 and even Geneva scored 10.8.

The first map is from the *Graphical-Statistical Atlas* (Federal Statistics Office, 1914) using data from 1901-10. The second is from the 100th anniversary edition of the Atlas, using the same districts and shading for data from 2001-10.

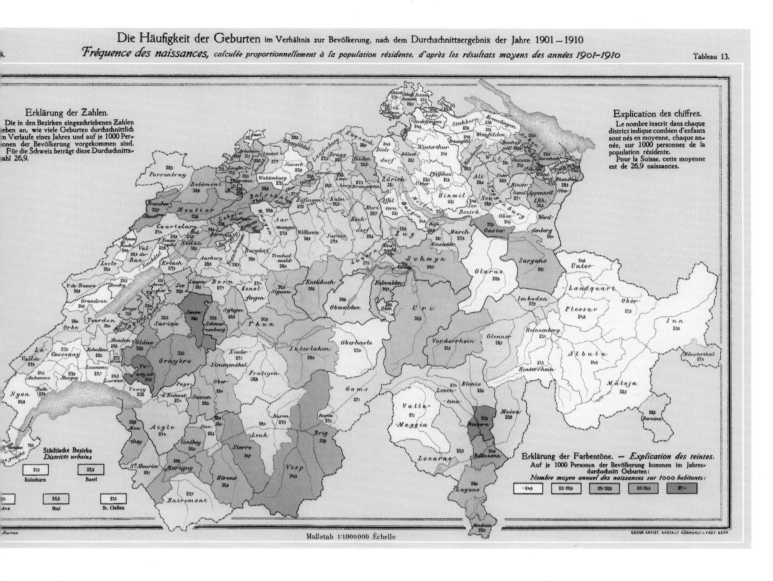

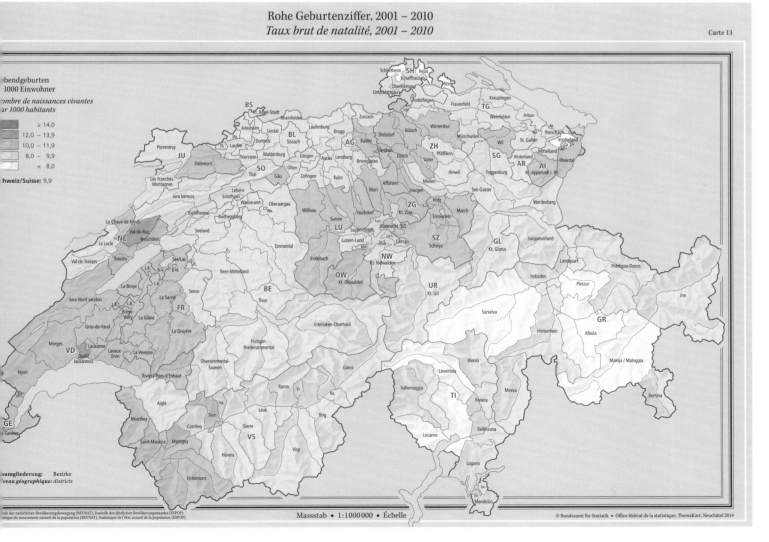

62   Counting the cows now and then

The animal that most symbolises Switzerland is perhaps the cow, preferably a brown one munching grass on a hill. And we know exactly where the cows live because the Swiss have counted them every five years for the past century.

These two maps are 100 years apart but both show the population density of cows, ie not absolute numbers but cows per km^2; the darker the shading, the denser the cows. The first map, from the 1911 census, shows Swiss cows mainly in their traditional home – the green foothills from Fribourg to Lake Constance, where most of the then 1.4 million cows lived.

The district with the highest density (165 cows per km^2) was Mittelland in Appenzell Ausserrhoden. It actually had far fewer cows than other districts, but is smaller in area. In contrast the mountain regions were largely empty, both in actual and density terms. Leaving aside Geneva (cow population 10 so a density of 0), the lowest figure was in Inn, Graubünden, with just 10 cows per km^2, far less than the national average of 62.

The second map is a century later and the difference is immediately clear. In 2011 more districts are shaded the darkest colour and cows are spread more evenly across the country. The Alpine areas are still the palest, but even there the density has increased.

The cow population in 2011 had grown to 1.6 million with a national average density of 106 per km^2. The district with the lowest density was now Maloja in Graubünden (14 per km^2) and the highest, Hochdorf in Canton Lucerne on 218.

The first map is from the *Graphical-Statistical Atlas* (Federal Statistics Office, 1914) using the livestock census of 21 April 1911. The second is from the 100th anniversary edition of the Atlas, using the same districts and shading scale for data from 2011.

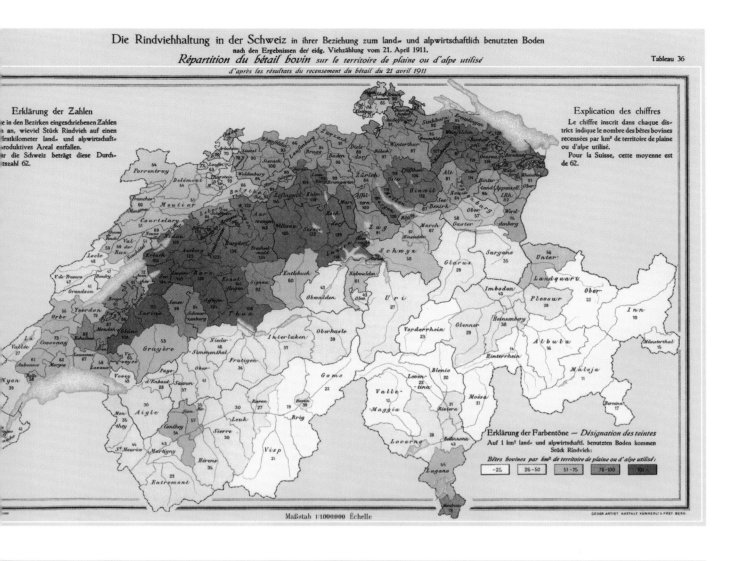

Die Rindviehhaltung in der Schweiz in ihrer Beziehung zum land- und alpwirtschaftlich benutzten Boden
nach den Ergebnissen der eidg. Viehzählung vom 21. April 1911.

Répartition du bétail bovin sur le territoire de plaine ou d'alpe utilisé

d'après les résultats du recensement du bétail du 21 avril 1911

Tableau 36

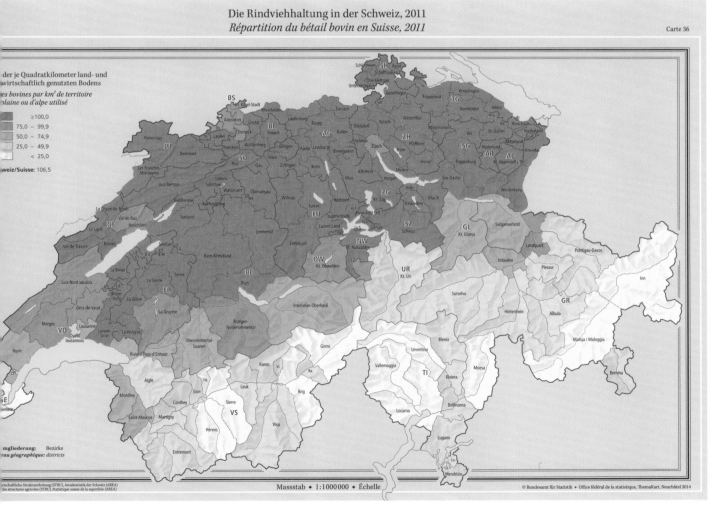

Die Rindviehhaltung in der Schweiz, 2011

Répartition du bétail bovin en Suisse, 2011

Carte 36

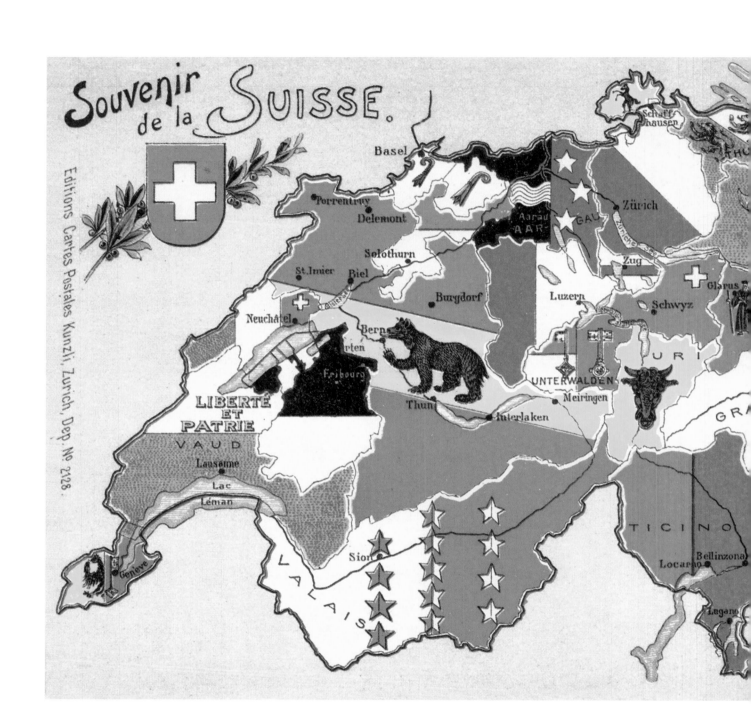

63   One nation of flags

The stars of Valais and the Bernese bear, the splashes of red and dashes of blue: some elements clearly stand out among this patchwork of 24 flags, each shaped to match its canton. Altogether they make an eye-catching souvenir postcard, this one dating from 1914.

Every canton has a square flag, once flown on the battlefield, now seen on balconies and town halls. Some pre-date the canton even being a canton, others were created specifically for its accession to the Confederation, eg Aargau in 1803. A few are variations on older emblems, such as the two lions of the medieval Kyburg dynasty in Thurgau.

Seven cantonal flags feature animals – a bear on Bern and both Appenzells, a bull on Uri, an eagle on Geneva, an ibex on Graubünden, the lions of Thurgau, and a ram on Schaffhausen. Glarus is now the only flag to feature a man, St. Fridolin, an Irish missionary who is believed to have converted the population to Christianity.

Graubünden's crest combines three coats of arms, one for each of the Three Leagues, as the canton used to be called. In 1932 the design was simplified, removing the men and stylising the three coats of arms into one. The flag not here is that of Jura, which became a canton only in 1979; that features a bishop's red crosier, symbolising the bishopric of Basel that once ruled the region.

The whole of the reverse side of the card is reserved for the address and stamp, as was normal in those days. Postcards were sent primarily for the picture and if a message was added, it had to be written in any space available on the front.

Written along the side of the postcard is "Editions Cartes Postales Kunzli Zurich", meaning that most likely it was produced by the Künzli brothers. Their postcard business ran from 1883 to 1965 and was famous throughout Europe in the early 20th century.

64   The land of cheese

For many people Swiss cheese simply means cheese with holes, and in many parts of the world holey cheese is called Swiss even when it's not. But not all cheese made in Switzerland has holes. In fact most of the 450-plus sorts of Swiss cheese are remarkably solid.

The big cheese in Switzerland is Emmentaler, with each round one metre in diameter and weighing up to 120kg (including the holes). Size aside, it had an important role in Swiss economic history. Cheese was traditionally made up in the mountains where the cows spent their summer holidays. Emmentaler was the first to be made down in the valleys all year round in village cheese dairies. That gave farmers a guaranteed winter market for their milk and created crucial economies of scale in cheese-making.

Across the *Röstigraben* from Emmentaler is its French-speaking cousin Gruyère, originating from the village with almost the same name (it has an –s on the end). But away from these two heavyweights are many others: Tilsiter, Appenzeller, Tête de Moine, Vacherin and countless *Alp-* and *Bergkäse* varieties. With so much on offer, it's no wonder the Swiss produce over 180,000 tonnes of cheese annually and eat 21kg of it per head per year.

You could spend ages looking at the colourful detail in this lively map, such as the three men sharing a fondue near Gruyères or the chef grating the abnormally large chunk of Sbrinz near Lucerne. As for the woman riding a cow while chasing a boar above Neuchâtel, perhaps it's better not to know any details.

Otto M. Müller (1913-2002) created this map in 1975 for the Swiss Cheese Union. It was one of many produced to show the variety of Swiss cheese but this has Müller's trademark style and calligraphy where the letters are rarely in a straight line.

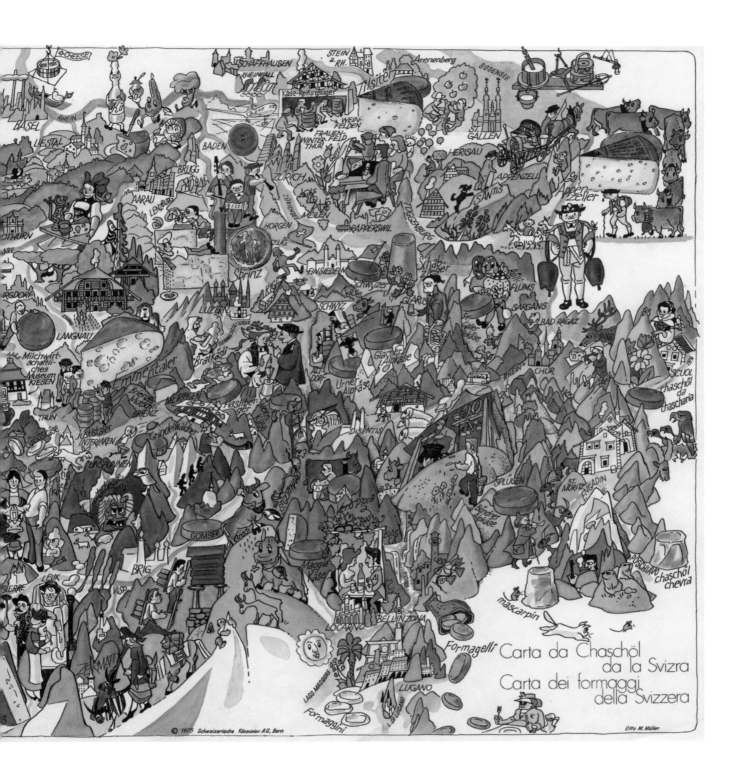

65   A political earthquake

Switzerland is at the heart of Europe but not at the heart of the European Union. Its membership has been out of the question ever since a historic referendum in December 1992.

The vote was actually on ratifying Swiss membership of the new European Economic Area (EEA), created by a treaty signed on 2 May 1992 between the 12 members of the European Community (EC) and seven members of EFTA, the European Free Trade Association. Switzerland had been a founding member of EFTA in 1960, but had slowly been moving closer to the EC; the Swiss government made the first moves towards applying for EC membership on 20 May.

Under the Swiss system of direct democracy, joining the EEA was subject to a referendum. In this crucial vote on 6 December the government, business leaders and unions were all in the Yes camp. The main opposition came from the right-wing SVP party under Christoph Blocher.

Turnout was unusually high (78.7%) and the result extraordinarily close. With over 3.5 million votes cast, the No side won with a majority of 23,836, or 50.3% of the vote. The majority was wafer-thin in terms of votes but, as this map shows, in terms of cantons, it was anything but. Of the 23 cantons (half-cantons count as halves in a referendum), only seven voted Yes, and the *Röstigraben* divide (see overleaf) between French- and German-speaking areas was clear.

After the vote, Switzerland stayed out of the EEA and suspended its negotiations for EC membership. Instead, a series of bilateral treaties were agreed (and approved by referendum) giving Switzerland access to the single market in return for implementing relevant legislation.

The front page from *Tages-Anzeiger*, a Zurich newspaper, on Monday 7 December 1992. All the French-speaking cantons, plus both Basels, said Yes to the EEA, with Neuchâtel the highest on 80%. The rest all said No, with Uri the highest on 74.9%.

Fr.

AZ 8021 Zürich, Montag, 7. Dezember 1992

Tages Anzeiger

: 01/248 41 21, Fax: 01/248 50 55
te: 01/248 41 11, Telex: 812 238, Fax: 01/248 41 91
nseratenannahme: 01/248 41 41

100. Jahrgang Nr. 285 Auflage 271 961
Unabhängige schweizerische Tageszeitung

Redaktion: 01/248 44 11
Telex: 812 236, Fax: 01/248 44 71
Werdstr. 21, 8004 Zürich. Briefe: Postfach, 8021 Zürich

ropa einigt sich – Schweiz gespalten

Massives Ja zum EWR in der Romandie, ebenso deutliches Nein in der Zentral- und Ostschweiz

apolitik ist das Schwei-
palten. Mit einem Zu-
on 50,3 Prozent lehnten
zer Stimmbürgerinnen
bürger am Wochenende
g zum Europäischen
raum ab. Eindeutig war
ehr: 16 Kantone stimm-
en EWR, sieben Stände
Deutschschweiz (ausser
das Tessin bodigten den
welsche Schweiz blieb
s in der Minderheit. Die
igung war mit 78,3 Pro-
ewöhnlich hoch.

LIENHARD, ZÜRICH

sich der Bundesrat, die
National- und Ständerat,
Parteien und sogar der
e Gewerbeverband für den
ausgesprochen. Trotzdem
mit einem knappen Ergeb-
am es auch, zumindest beim
Mit 1 786 121 Nein gegen
haben die Schweizer die
n EWR verworfen. Wegen
hrs stand aber die Schwei-
g des EWR-Beitritts schon
nntagnachmittag fest: Aus
n Schweiz – und aus dem
en lauter Nein. Gegen diese
on 16 Kantonen hatten die
d Basel mit ihren sieben
en keine Chance.

zent Nein
Prozent Nein

n Anteile von Nein-Stim-
kleine Innerschweizer und
Stände: Uri mit 74,5 Pro-
r auch Schwyz, Obwalden
l Innerrhoden mit mehr als
Nein. Die übrigen Inner-
l Ostschweizer Kantone so-
bewegen sich zwischen 60
ent Nein-Stimmen. Etwas
cher, aber doch noch dage-
ich die Wirtschaftskantone

Beitritt der Schweiz zum EWR

JA
50–60%
über 60%

NEIN
50–60%
über 60%

TA-GRAFIK

Zürich und Zug sowie die grossen Mittel-
land-Kantone Aargau, Solothurn und
Bern mit Nein-Stimmen-Anteilen von
60,1 Prozent und weniger. Im Kanton Zürich war das Resultat am knappsten: Das
Nein in der Landschaft wog ein bisschen
schwerer als das Ja in der Stadt. End-
ergebnis: 51,5 Prozent Nein.

Neuenburg: 80,1 Prozent Ja

Die deutsche Schweiz lehnte den EWR
wuchtiger ab als erwartet. Spiegelbild-
lich war das Ja der französischen
Schweiz deutlicher als vorausgesehen:
Der Europa-Graben zwischen den Lan-

desteilen ist tief und veranlasste die Romandie zu äusserst besorgten Kommen-
taren. In den Kantonen Jura, Genf, Waadt
und Neuenburg stimmten 77,2 bis 80,1
Prozent ja zum EWR. Die beiden Kantone mit Deutschschweizer Minderheiten, Freiburg und Wallis, lieferten im-
merhin noch 64,9 und 55,9 Prozent Ja.

Die vielen Ja aus der Romandie hätten
beinahe ausgereicht, um ein gesamt-
schweizerisches Volksmehr zu errei-
chen. Aber die Übermacht der 16 Nein-
Stände war am Schluss doch vorhanden –
wenn auch nur mit einem Unterschied
von 23 105 Stimmen.

«Schwarzer Sonntag»

In einer ersten Reaktion sprach Bundesrat Delamuraz von einem «schwarzen
Sonntag» für die Wirtschaft, für die Ar-
beitnehmer und für die Jugend. Und auch
am Bildschirm zeigte sich der Unterschied zwischen Deutschschweiz und
Romandie: Wer die Kommentare der drei
Bundesräte Felber, Delamuraz und Kol-
ler ab 18 Uhr live am Fernsehen verfol-
gen wollte, musste die welsche SRG wäh-
len. Das Fernsehen der deutschen und rä-
toromanischen Schweiz sendete das
Gutenacht-Gschichtli «Pingu als Küchen-
chef».

CHAPPATTE

66   Across the Röstigraben

Every country has its own regional divisions, be
they between north and south or town and country, but
Switzerland has a very distinctive one. So distinct that
it has its own name: the *Röstigraben*.

Literally meaning "fried-potato trench", it refers
to the German-speaking Swiss loving *Rösti*, a dish of
grated fried potatoes. But it has nothing to do with
potatoes and there is no trench; the *Röstigraben* is in
fact an invisible barrier between the German- and
French-speaking areas. Switzerland has four national
languages (see overleaf) but it is this German-French
division that features most prominently in the Swiss
psyche.

Simply because *Röstigraben* doesn't only refer to a
linguistic divide but is also shorthand for a political one,
with the French-speaking Romandie generally being
more left-leaning and pro-European than the rest of the
country. Many referendum results clearly show this
division (see previous page), so that the Suisse Romands
sometimes have been called less patriotic and less
Swiss.

Some also like to see it as a cultural gap between
west and east, often portrayed in stereotypes:
The *Rösti*-eaters in the east are arrogant perfectionists
who ignore the rest of the country; the fondue-eaters in
the west are disorganised layabouts who refuse to
speak German.

But as this Chappatte cartoon from the 1990s
shows so cleverly, on both sides of the divide (be that
a ditch or a hedge), the people are remarkably similar.
They might not speak the same language or have
the same views, but they are both Swiss. And want to
stay Swiss.

Patrick Chappatte (born 1967) is a Lebanese-Swiss cartoonist born in Pakistan and
raised in Singapore and Switzerland. He now lives in Los Angeles and his cartoons appear
in *Le Temps* in Geneva, the *NZZ am Sonntag* in Zurich and the *International New York Times*.

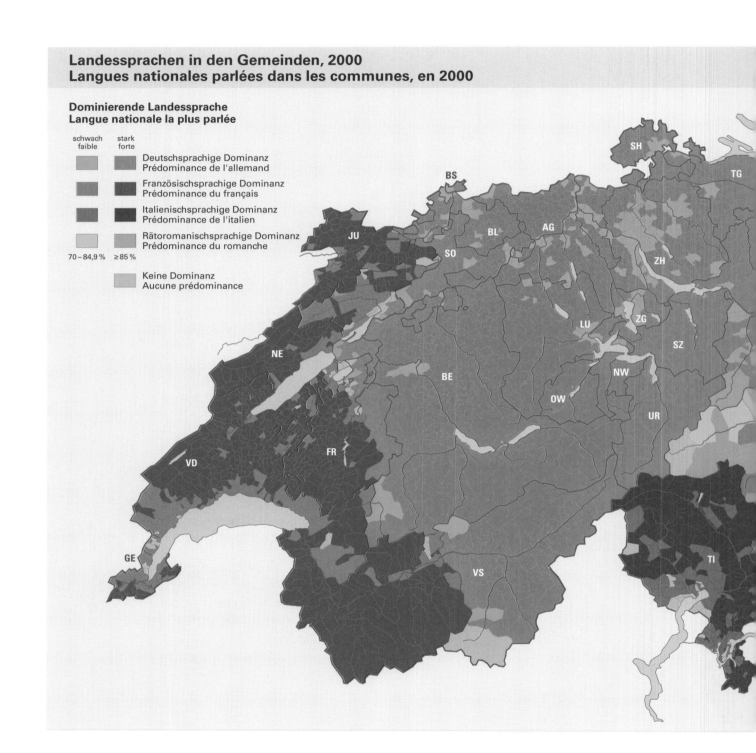

Landessprachen in den Gemeinden, 2000
Langues nationales parlées dans les communes, en 2000

Dominierende Landessprache
Langue nationale la plus parlée

schwach stark
faible   forte

Deutschsprachige Dominanz
Prédominance de l'allemand

Französischsprachige Dominanz
Prédominance du français

Italienischsprachige Dominanz
Prédominance de l'italien

Rätoromanischsprachige Dominanz
Prédominance du romanche

70–84,9 % ≥ 85 %

Keine Dominanz
Aucune prédominance

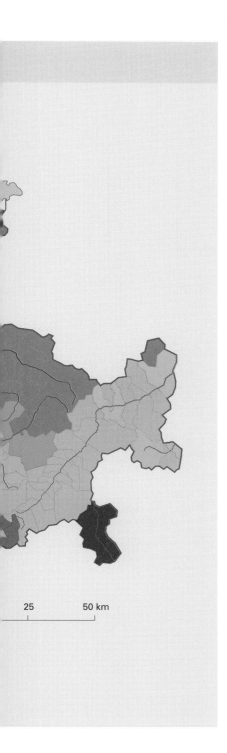

25   50 km

67   One country but four languages

Schweiz, Suisse, Svizzera, Svizra. In its four national languages Switzerland has four different names though they are rarely all used in one sentence. And while many Swiss people are bi- or multilingual, there are stark dividing lines between the language areas, as this map from the year 2000 dramatically illustrates.

Every community is shown with its main language spoken: green for French, red for German, blue for Italian and orange for Romansh. Bold colours are communities with over 85% of the population speaking that language, paler shades for 70 to 84.9%. The latter are very noticeable along Lake Geneva and in cities like Zurich and Zug, home to large numbers of foreigners. Grey is where a community has no single language as the dominant one.

The clear German-French division is known as the *Röstigraben* (see previous page). Some communities along it are paler shades or in grey, but quite often the language switches simply by crossing the line. Usually there isn't even a mountain pass separating the two, unlike the change from German in the north to Italian in the south.

Of the four national languages, German is spoken by 64% of the population and French by 20%. Three cantons are officially bilingual: Bern, Fribourg and Valais, all straddling the *Röstigraben*. Italian constitutes only 6.8% nationally but is the main language in Ticino and southern Graubünden.

A referendum in 1938 made Romansh the fourth Swiss national language, although it is not an official language of the state, and it accounts for a mere 0.5% of the population, nearly all of them in tri-lingual Graubünden.

This Bundesamt für Statistik map is a snapshot of linguistic Switzerland at the turn of the millennium, based on the 2000 census. Since 2010 people could respond as speaking more than one language instead of just one permissible until then.

68   Colours of the political spectrum

Referendum results don't just decide the issues at hand, they also give an insight into the political shape of the country, based not on elections every four years but on votes about specific issues every three months.

The map was created using 184 national referendum results between 1982 and 2002. While politics dictates the map's shape, its detail is down to the population. The colours are the three main languages – green for German, pink for French, yellow for Italian – and contours represent the size of each community: the higher the hill, the larger the population.

The horizontal axis runs from left to right, both in real and political terms, and is based on three main referendum themes: social policies, civil rights and defence issues. The vertical axis, from liberal at the top to conservative at the bottom, also covers three themes: foreign affairs, immigration and integration, and institutional reform.

It's immediately clear that most German-speaking communities are more right-wing than the rest of the country but fairly evenly split between liberal and conservative. In contrast the French-speaking region is almost all liberal and left, with Jura's capital Delémont the most extreme. Ticino veers left but straddles the middle of the other axis.

There's also a definite urban-rural divide, with the five biggest Swiss cities all in the left-liberal quadrant, Geneva (Genf) being the most left. Directly opposite, ie right-conservative, are mainly rural communities, particularly in central and eastern Switzerland, with Unteriberg in Canton Schwyz as conservative as it gets.

The map comes from the book *Atlas der politischen Landschaften* by Michael Hermann and Heiri Leuthold, published in 2003 by vdf Hochschulverlag an der ETH Zürich. The book also includes political maps of every canton in the same style.

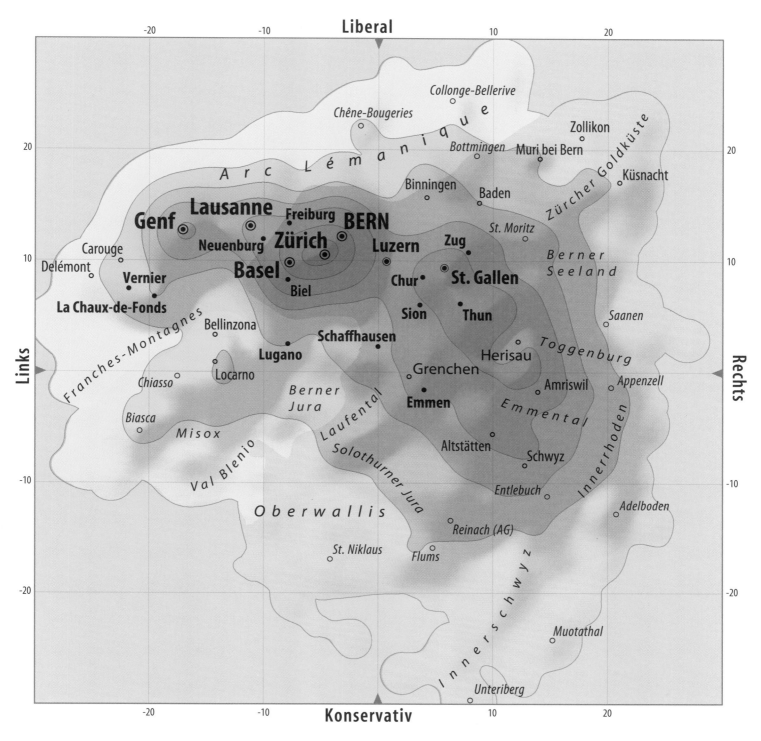

Liberal

-20   -10   10   20

20

Collonge-Bellerive

Chêne-Bougeries

Arc Lémanique

Zollikon

Bottmingen   Muri bei Bern

Binningen   Baden   Küsnacht

Zürcher Goldküste

Lausanne   **Freiburg**   **BERN**

St. Moritz

Genf   **Neuenburg**   **Zürich**   **Luzern**   Zug   *Berner Seeland*

Carouge   **Basel**   Biel   **Chur**   **St. Gallen**

Delémont   **Vernier**   Sion   **Thun**   *Saanen*

La Chaux-de-Fonds   Bellinzona   *Franches-Montagnes*   **Schaffhausen**   Herisau   *Toggenburg*

Links   **Lugano**   Grenchen   Amriswil   *Appenzell*   **Rechts**

Chiasso   Locarno   *Berner Jura*   **Emmen**   *Emmental*   *Innerrhoden*

Biasca   *Laufental*   Schwyz

Misox   *Solothurner Jura*   Altstätten

Val Blenio   Entlebuch   Adelboden

-10

Oberwallis   Reinach (AG)

St. Niklaus   Flums

Innerschwyz

-20

Muotathal

Unteriberg

Konservativ

-20   -10   10   20

© Michael Hermann & Heiri Leuthold, sotomo

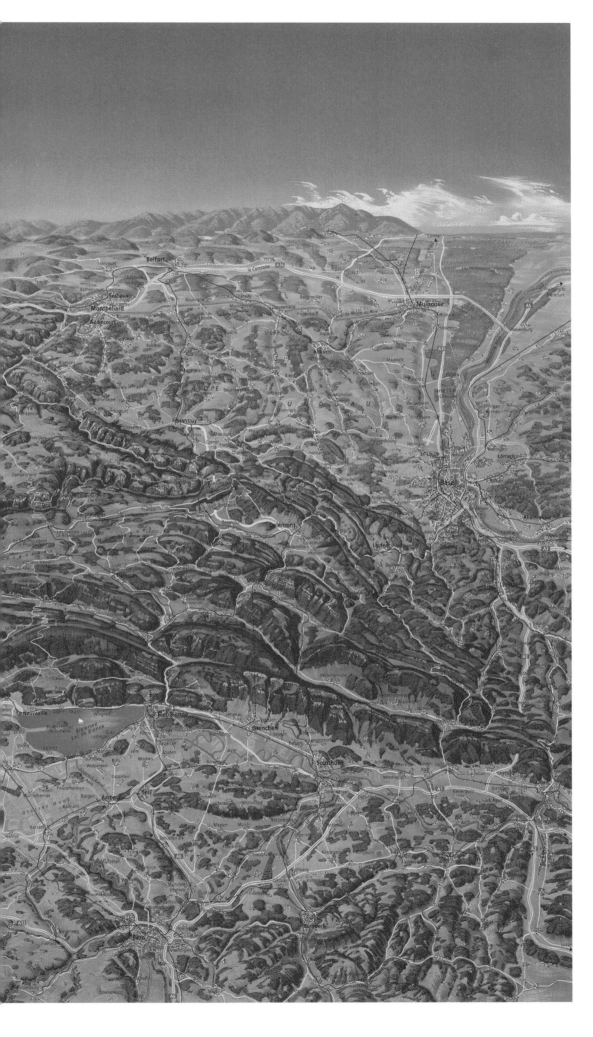

69

A clockwork country

Watches are an iconic Swiss product so it's no wonder that the region where most of them are produced has been christened Watch Valley.

▶

69   A clockwork country

◄

Watches are an iconic Swiss product so it's no wonder that the region where most of them are produced has been christened Watch Valley. It's as much about marketing as manufacturing, as this tourist map of the area proves.

The Swiss have Jean Calvin, the Protestant preacher of Geneva, largely to thank for their watch industry. He welcomed Hugenot refugees who were fleeing Catholic France, with many watchmakers among them. But he also banned fancy jewellery in 1541 so that the skilled craftsmen of Geneva had to switch from creating trinkets to fashioning timepieces.

The world's first Watchmakers' Guild started in Geneva in 1601 but the city soon wasn't big enough for the industry, which spread out north and east into the hills between Lake Neuchâtel and the French border. For the past century, 90% of Swiss watchmaking has been concentrated in this Watch Valley, where famous brands such as Omega, Tag Heuer and Tissot are based.

Watches are a big income-earner for Switzerland, with annual exports totalling 28 million timepieces and 20.6 billion Swiss francs. It isn't just the watch companies that benefit but the small firms supplying the intricate parts. Many of those are also located across Watch Valley.

The two towns at the heart of the region, both geographically and historically, are La Chaux-de-Fonds and Le Locle, which were jointly designated Unesco World Heritage sites in 2009. After disastrous fires (see page 80) both towns were planned with the watchmakers' needs in mind so that working space and living space co-existed perfectly.

A 2004 map from Neuchâtel Tourism with a northwest orientation and a bird's-eye perspective. Watch Valley stretches along the Three Lakes region, named after Lake Neuchâtel (the largest on this map), Lake Biel and Lake Murten (the smallest).

100,000
inhabitants

70　The real shape of the population

A normal map of Switzerland shows the 26 cantons as they are geographically, from the largest – Graubünden at 7,105.2 km² – down to the smallest, Basel-Stadt at merely 37.1 km². But what if that map showed the cantons based not on land area (see Appendix) but on population? Just how different would it look?

Very different is the answer, as this example shows. Each canton now represents its population rather than area, skewing the shapes and sizes of most of them. While it's clear that Zurich is the most populous canton, with 1.4 million inhabitants, the biggest distortions appear in the most and least densely populated areas. Geneva and Basel-Stadt both swell into giant tumours on the edges of the country, while Uri and Graubünden shrivel down to anorexic proportions.

Some cantons look almost the same. Vaud is large both in terms of population and area, ranked 3rd and 4th respectively, so barely changes; at the other end of the scale tiny Appenzell Innerrhoden also remains constant as it is ranked last for population (just 15,700) and 25th in area.

Only two cantons achieve the same ranking in area and population. Out of 26 cantons, Nidwalden is 22nd and Bern 2nd in both sets of data, so that in this map their shape is determined more by the cantons around them growing or shrinking.

Over 58% of Switzerland's eight million inhabitants live in six cantons - Zurich, Bern, Vaud, Aargau, St Gallen & Geneva – while the thirteen smallest cantons in terms of population account for just 15% of the total.

The map was created by a man from Canton Zurich calling himself "aetheist Raskalnikow". Using a computer programme, he skewed a normal Swiss cantonal map (see Appendix: The Swiss cantons on page 218) with official population data.

71   A very Swiss supermarket

It all began with five trucks that sold six products at 178 locations around Zurich. That was in August 1925 and 90 years later Migros has grown into Switzerland's largest retailer and largest employer. So big, in fact, that it has its own regional structure.

As this map shows, Migros is organised into ten regions that don't always follow cantonal or language boundaries: Zurich (in pink) loses the eastern half of its canton but gains Glarus, while German-speaking Basel (yellow) extends into French-speaking Jura. The vast Ostschweiz region (grey) covers parts of seven cantons, but Geneva (purple) is all on its own.

The Aare region (blue) encompasses the cantons of Bern, Aargau and Solothurn, and is the one with the most members: almost a quarter of the national total of 2.1 million. The Migros members are mostly the customers themselves, with each member belonging to a regional *Genossenschaft*, or co-operative.

These ten co-operatives together form the single entity known as Migros, which was transformed in 1941 from a limited company into a co-operative for its customers. Since then, the original nine regions have grown organically, becoming ten as Migros expanded, but also merging and moving regional boundaries over the decades.

The man behind the big orange M was Gottlieb Duttweiler, a Zurich businessman who decided to cut out the middleman and sell directly to the public. Alongside supermarkets, came a travel agency, petrol stations, restaurants and a bank. And 1% of the profits is ploughed into cultural activities, such as adult education schools.

Map provided by the Migros Club Schools, founded in 1944 as a language school. Courses expanded to include artistic pursuits, hobbies, and sports, and the 50 Club Schools now constitute Switzerland's largest adult education centre.

Fantasy Switzerland

Things that never happened and alternative views of the country

72 Welcome to Henripolis!

Look at a map of Switzerland today and there is one town that will not be there: Henripolis. It isn't on the map because it was never built; but it was planned and mapped out on the shores of Lake Neuchâtel.

In 1625 most of Europe was seven years into what would become known as the Thirty Years' War, the big religious conflict of the 17th century. At the heart of this turmoil was the Swiss Confederation, and one of its allies - the principality of Neuchâtel, not yet Swiss but tied to the cantons for many years (see page 74).

Neuchâtel was then a Protestant principality ruled by a French Catholic prince, Henri II of Orléans-Longueville (1595-1663), who rarely visited his realm and didn't get on too well with his citizens. As part of this power struggle between prince and people, and to weaken Neuchâtel's ties with neighbouring Canton Bern, Henri planned a new city.

On 24 June 1625 Henri signed the founding charter for what he very modestly had christened Henripolis. It would be built beside Lake Neuchâtel between the communities of Marin and Epagnier, near the River Zihl: on this map it's the red and white grid at the right-hand end of the lake, less than 10km from the spires of Neuchâtel itself.

Of course this wasn't only about princely vanity and power politics, it was also about money and trade. Building a commercial safe haven in the peaceful stable centre of a war-torn continent could work, especially if that was linked to the main waterways of Europe.

Henripolis lay on a direct route to the North Sea via the Zihl, Lake Biel, the Aare, and the Rhine. A day's land journey separated Lake Neuchâtel from Lake Geneva and the Rhone (and so the Mediterranean), but even that missing link would soon be closed by the planned Canal d'Entreroches between the two lakes.

▶

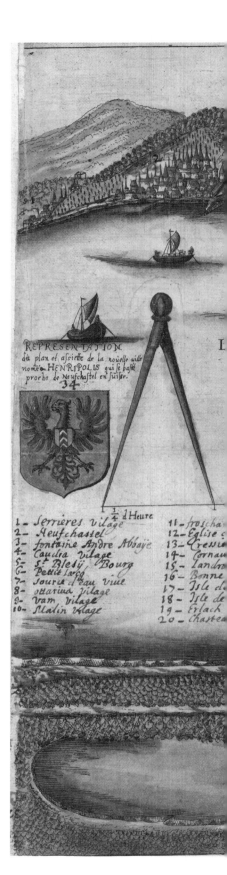

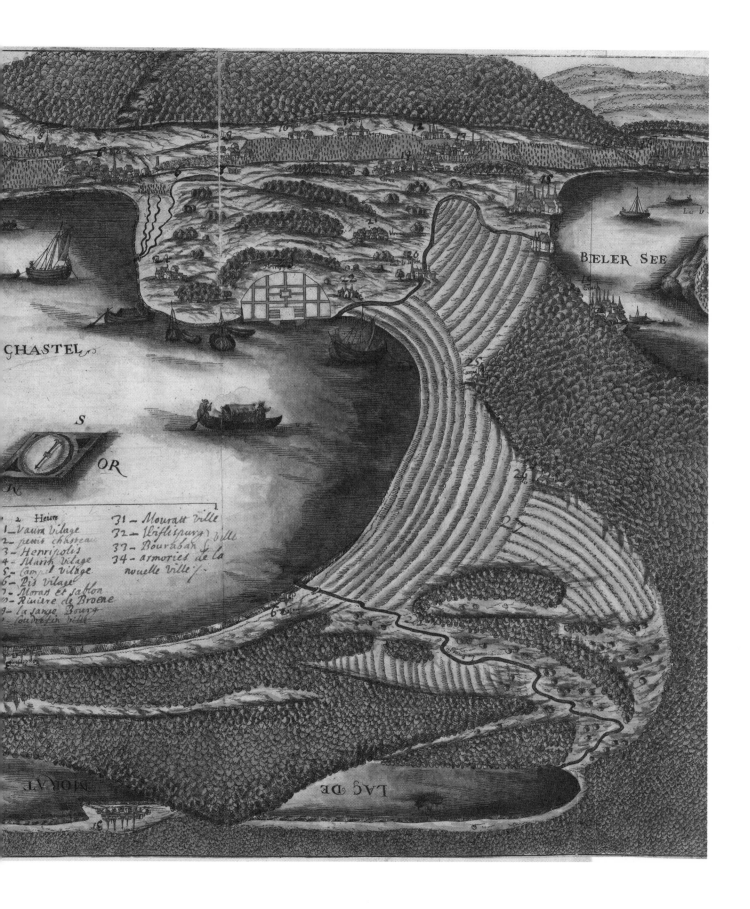

73 ◄

With such a route the Protestants traders of the northern Europe, such as the Dutch East India Company, could reach the Mediterranean without going near Catholic Spain or France.

The city was planned on a site of 54 hectares of land divided into 1,650 plots, as this detailed map shows. It was to be built in a polygonal half-moon shape with streets on a rigid grid pattern. The north-south axis would run from the palace (1 on the map) down to the port (7), with its market hall (5) and granary (6); the other would run between the two main churches (both 4), one German-speaking, one French- but both Protestant. At the epicentre was the town hall (2). At every main intersection was a fountain, aesthetically pleasing and hygienically necessary.

This model geometric city had enough space for 13,000 residents (Bern then had only 10,000 inhabitants), who would be entitled to commercial and trade privileges as well as religious freedom and existing political rights. The prospectus of 1626 made it sound like a political Utopia as well as an Eden with fish from the lake, good white wine and all manner of fruit waiting to be picked.

But it was never built. Quite aside from not raising enough capital to finance the project, it proved impossible to find enough landowners willing to sell land or enough potential new inhabitants. Plus neither Neuchâtel, nor its main Swiss ally and protector Bern, wanted to support the creation of a potential rival. Henripolis is the city that never was.

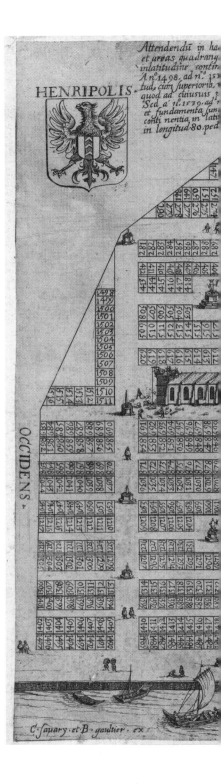

Both maps formed part of the official prospectus, published in 1626 in three different language editions: French, German and Dutch. This city map came from the French one, published in Lyon by Claude Savary and Barthélemy Gaultier.

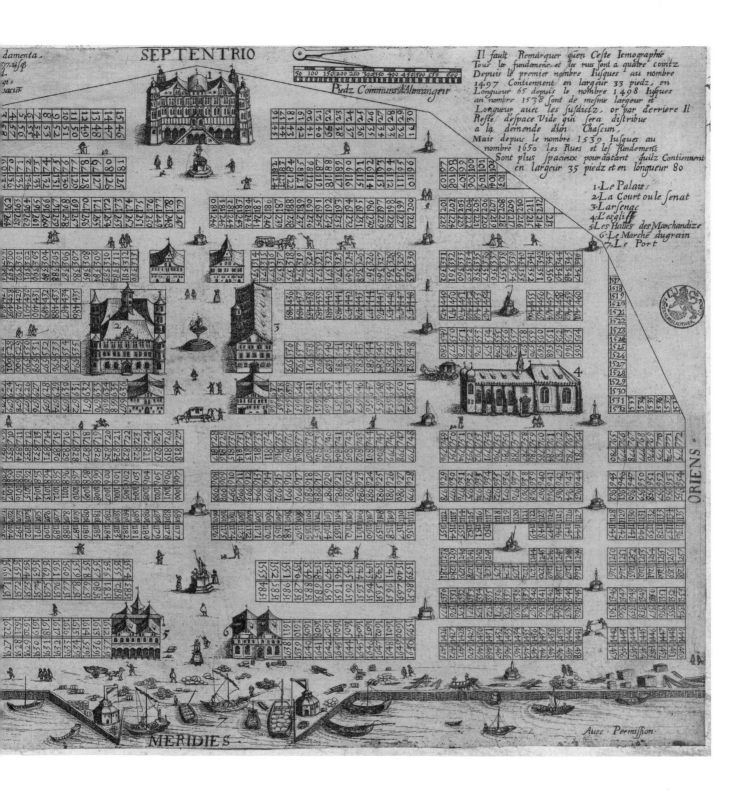

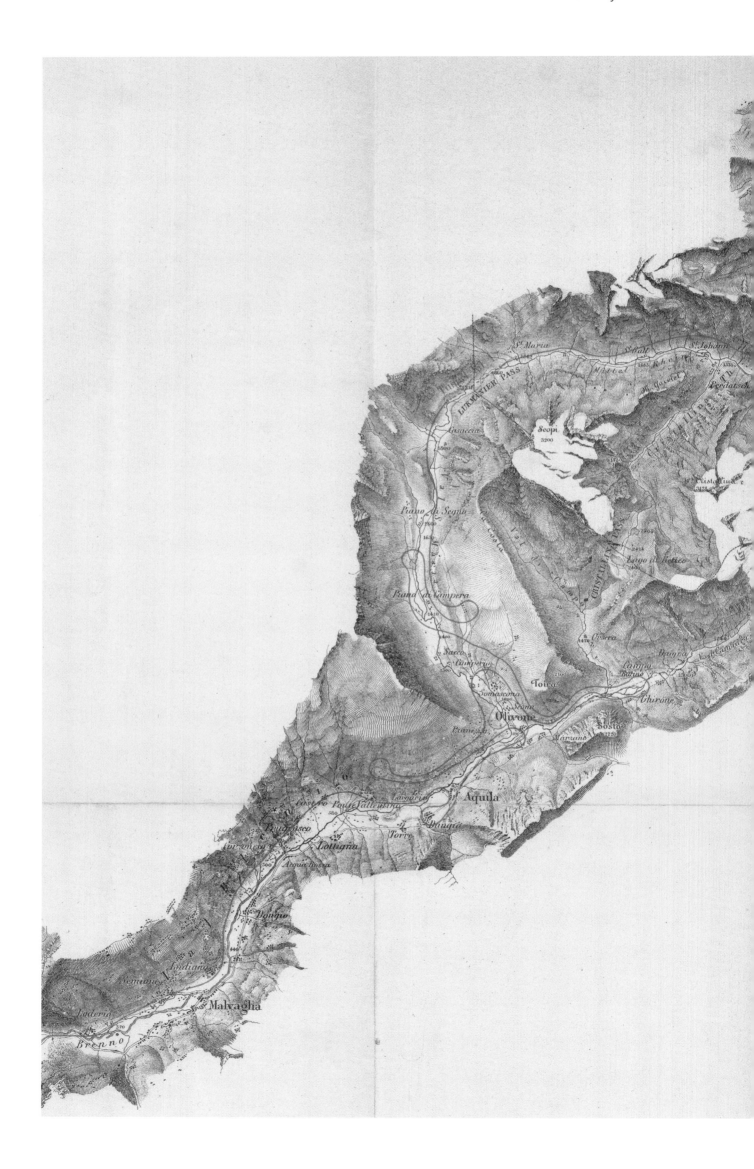

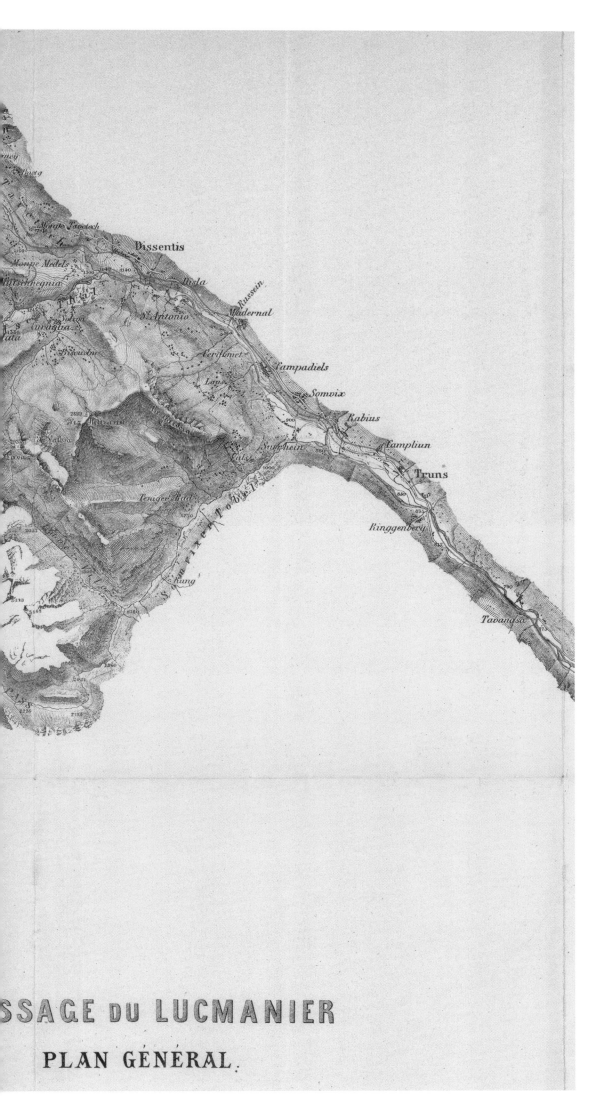

74

A railway to rival the Gotthard

It could have been the pinnacle of Swiss engineering in the 19th century.
It would have been built instead of the Gotthard tunnel.
It should have been a name that every train lover knows.

▶

74   A railway to rival the Gotthard

◄

It could have been the pinnacle of Swiss engineering in the 19th century. It would have been built instead of the Gotthard tunnel. It should have been a name that every train lover knows. But Lukmanier is simply a footnote in Swiss railway history: the train line that never left the station.

Sitting at 1915 m above sea level, the Lukmanier Pass links Disentis in Graubünden 20km to the north with Olivone in Ticino 19km to the south. Running roughly parallel with the Gotthard Pass to the west, it was an important route into Italy in the Middle Ages, largely down to its relatively low altitude, 200 m below the Gotthard.

By the mid 19th century the Lukmanier route was a rural byway, overtaken by developments elsewhere, but then became a contender for the new north-south railway. Starting in Chur, the proposed line ran along the Rhine valley to Disentis, and then clambered up into the mountains.

After a 1.7km tunnel under the Lukmanier Pass itself, the line descended in a series of s-bends, with a maximum gradient of 25%, to Olivone and Biasca. So after climbing from 1150m at Disentis up to an altitude of 1865m, it then dropped to Biasca at only 289m. Plan B had a much longer tunnel, shown by the dotted line on the map, under Piz Greina to bypass the mountain section. At 17.4km it was 2km longer than the rival Gotthard tunnel and proved too expensive to be pursued.

In 1869 the Swiss government chose the Gotthard for the new railway through the Alps, so instead a new road was built over the Lukmanier pass in 1872-7.

This map from the 1860s is oriented to the west with Graubünden on the right and Ticino on the left. It was printed by Wurster, Randegger & Co in Winterthur. Founded in 1842, the company went bankrupt in 1924 and was bought by Orell Füssli.

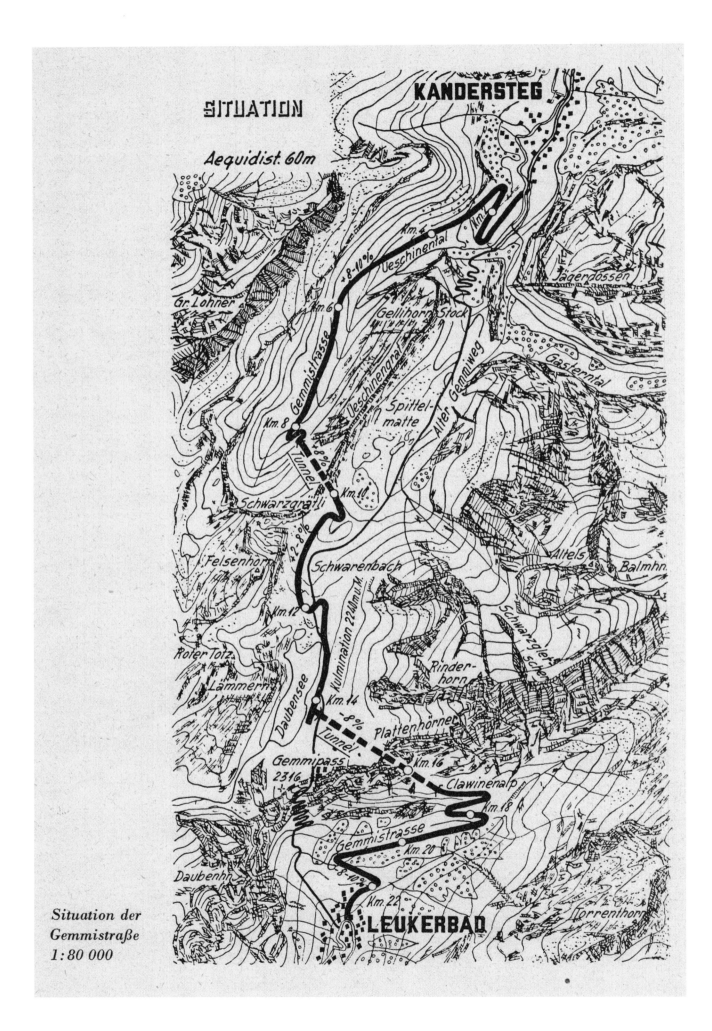

Situation der
Gemmistraße
1 : 80 000

75   A road to nowhere

Die Gemmistraße im Üschinental mit Alpbachviadukt und Üschinengrattunnel

The cantons of Bern and Valais share a long border so it's only natural that the Swiss wanted to build a road between the two. One big problem: mountains. They don't usually stop the Swiss (otherwise they would build few roads) but they do present logistical and financial obstacles.

In 1948 a motorway was planned over the Gemmi Pass between Kandersteg in Canton Bern and Leukerbad in Valais. It was one of the shortest and cheapest routes between north and south, would cut the travelling distance by 115km, and could stay open all year round.

The road was to be 23km long and six metres wide, with its highest point at Daubensee, 2240m above sea level. An ambitious tunnel would then take the road down under the Gemmiwand, a 900m-high cliff behind Leukerbad. The toll of two francs per person would help offset the construction costs and supply a projected annual income of 2 million francs, based on summertime estimates of 800 cars, 40 coaches and 200 motorbikes per day.

A Gemmi road had been planned before, such as in 1840 when the Valais parliament approved one project while the route was being measured. But it never happened and the opening of the Lötschberg rail tunnel in 1913 put any other ideas on ice for three decades. By 1949 over 210,000 cars a year were using the Lötschberg auto-trains and a road seemed a viable alternative.

This plan was published in 1952 but died six years later when the federal motorway plans (see page 142) chose the Rawil Pass over Gemmi for a Bern-Valais road. As it turned out neither was built and a footpath remains the only way over the Gemmi.

The map and illustration were taken from the booklet *Eine Autostrasse über den Gemmipass von Kandersteg nach Leukerbad* (1952), published by the committee in favour of building a motorway over the Gemmi Pass.

76 Zurich's missing metro

Switzerland's largest city once had dreams of being a real metropolis complete with a brand new underground system. It was going to be the dawn of a new high-tech age. It would transform Zurich's streets. It never happened.

In the early 1970s plans were drawn up for an underground railway (or *U-Bahn*) to help reduce Zurich's congestion. The first line - 26.1km long, half of it underground - would run from Kloten and the airport through the city centre and out to Dietikon. There were to be 30 stations with the principal interchange under Zurich's main train station. Further lines were envisaged out to Baden and Thalwil.

Construction time was forecast to be 13 years, with the first section out to the airport opening in 1980. It was predicted that by the year 2000 travel time between the airport and Zurich main station would be 17 minutes by *U-Bahn* (by the way, it is now 9 minutes by train or 48 minutes by tram).

But first it had to be approved by the voters because the price tag of 1.8 billion Swiss francs was being shared three ways between the city, the canton and the federal government. Opponents supported public transport as an ideal but criticised the endless push for growth and its effects on both the quality of life and the environment. Many simply didn't want Zurich to be a world city.

Polls showed a clear Yes but in the end the vote wasn't even close. On 20 May 1973 the No vote was 71% in Zurich, and 57% in the canton. The traffic chaos remained and Lausanne became home to Switzerland's first metro (see page 150).

A map of Zurich's phantom metro. The 2km Milchbuck Tunnel to Schwamendingen was built in 1971 as a test; today it is used by trams 7 and 9, with three underground stations at Tierspital, Waldgarten and Schörlistrasse.

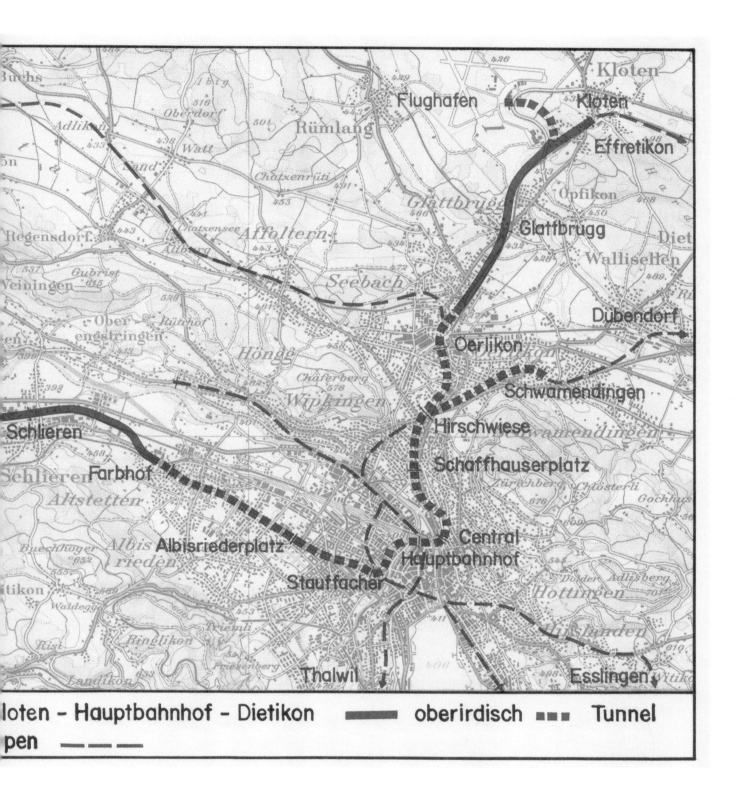

loten - **Hauptbahnhof** - Dietikon ▬▬▬ **oberirdisch** ▬▬▬ **Tunnel**

pen ▬ ▬ ▬

77   The cantons of tomorrow?

Most of the 26 Swiss cantons have borders shaped by accidents of history. Many have been around for centuries, nine were created in 1803 or 1815 and one, Jura, is still relatively new - it only became a canton in 1979.

But cantonal boundaries are not set in stone and although no cantons have ever merged, it would be possible. It already happens at a community level: in 1990 there were 3,021 Swiss communities, in 2014 only 2,352. Voters have consistently rejected proposals to fuse cantons, such as reuniting the two Basels or merging Geneva and Vaud. But it isn't unthinkable.

Merging cantons would iron out border anomalies and could be much more efficient, both politically and economically. Over 700 inter-cantonal agreements are currently in effect while cantonal disagreements can delay policies for years. It's a question of logic over history: Appenzell Innerrhoden has fewer than 16,000 inhabitants but should it exist purely because of history?

Pierre-Alain Rumley proposed a new Switzerland of 13 cantons, just as it used to have for over two centuries (see page 20). Seven cantons would stay almost the same, mainly due to geography or language, but 17 cantons would become five. Each has a metropolitan area at its heart – Zurich, St. Gallen, Basel, Lucerne, Bern – and includes all or parts of neighbouring smaller cantons, which would disappear completely.

One totally new canton – the 13th - would be created. 'Arc jurassien' would bring together Jura, Neuchâtel and the French-speaking Bernese Jura in a new canton of 340,000 people. It's a radical plan that might be put to the vote one day.

Pierre-Alain Rumley (born 1950) is from Neuchâtel University and a former director of the Federal Office of Spatial Development. This map comes from his book *La Suisse demain* (or *Switzerland tomorrow*) from 2010, which proposed a Switzerland with 9 or 13 cantons.

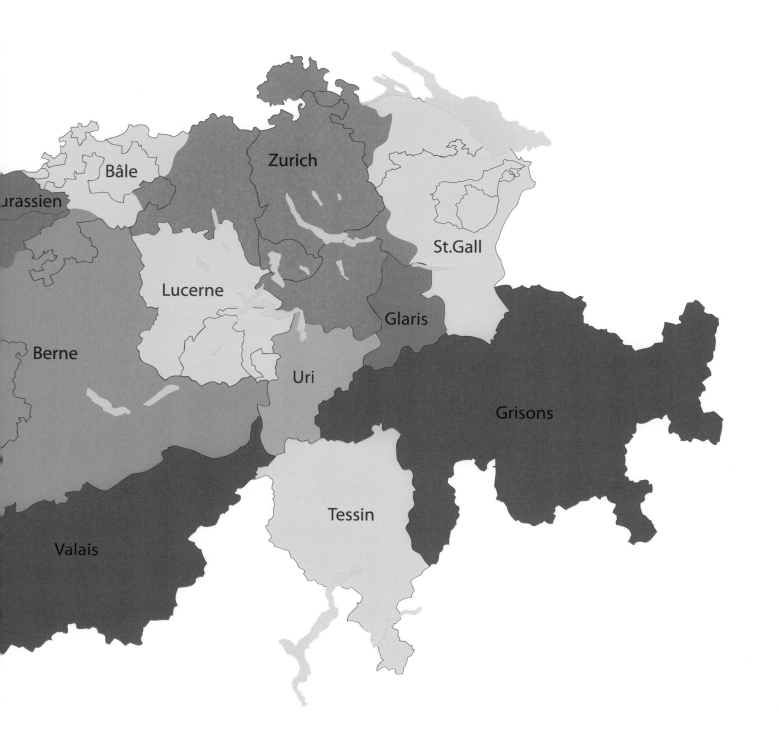

78   The underground intercity

Bern to Zurich in 12 minutes; it takes that long to cook spaghetti let alone travel 104km. Today that journey needs 56 minutes even with a normal high-speed line, while Geneva to St. Gallen needs four hours, but soon that longer trip could be possible in under an hour.

It's straight out of a sci-fi movie set at a time when electric cars fly and sleek trains speed silently past. But this is not the latest Hollywood blockbuster, this is a real plan for an underground train travelling at super high speed. It is called Swissmetro.

Trains would travel at over 500 km/h in two tunnels 50-100 m underground using maglev and vactrain technology. In other words, magnetic levitation propels the wheel-less trains through vacuum tunnels where the atmosphere is regulated to reduce air resistance. Air pressure inside the train would be similar to inside a plane flying at maximum 15,000 metres. Trains could run every six minutes in rush hour, thus simultaneously tackling the overcrowding on Swiss roads and rails.

This map shows the initial three lines: the main pilot project from Bern to Zurich, plus Geneva to Lausanne and Zurich airport to Basel airport. Later lines would link these and extend the network to St. Gallen, Chur and Sion, as well as joining up with an envisaged Eurometro to Lyon or Munich.

Until now the project has failed to gain ground, despite its ecological advantages and not being subject to problems caused by weather and cows on the line. But maybe in the future we'll all be going underground.

Tronçons Prioritaires
Priority Stretches

- Bern – Zürich (104 k
  - ● Test Reiden (17 km)
- Genève – Lausanne
  - ● Test Céligny (18 km)
- Zürich Flughafen – B
  - ● Test Dielsdorf (11 km

V-S-H- ○ *SWISSMETRO* Te
Test Gallery Hagerba

Tronçons Ultérieurs
Secondary Stretches

- Bern – Lausanne (81
- Zürich – Winterthur – St. Gallen (79 km)
- Zürich – Chur (93 km
- Lausanne – Sion (77 km)
- Rail 2000
- Connection AlpTransit-Stretches
- Eurometro

Lyon

Pro Swissmetro is a lobbying group trying to get this futuristic rail project back on track. It took over from Swissmetro AG, the company that went into liquidation in 2009 after failing to get the necessary political and financial support.

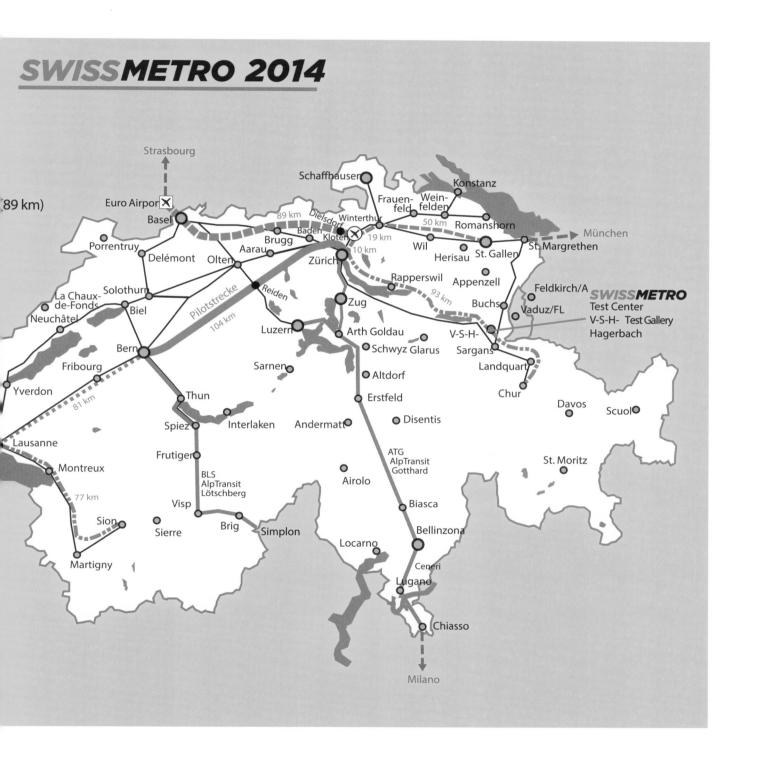

SWISSMETRO 2014

79   Greater Switzerland with 40 cantons

The shape of Switzerland has not changed since its international borders were set in 1815 (see page 44). But what if that were different? What if those regions that had once been Swiss had stayed Swiss? What if we included all of them, which are now in other countries, in a Swiss map? What would this Greater Switzerland look like?

Before Switzerland gave up its empire-building in the 16th century, it controlled many neighbouring regions, some only for a few years as effective colonies, others for centuries as close allies under Swiss protection.

All were re-conquered, traded away or forcibly taken from the Swiss but here they have been reinstated, along with some parts that wanted to be Swiss but couldn't. The individual maps opposite show these 15 neighbouring areas, starting in the far east and going clockwise around the Swiss border. Each has correspondingly numbered notes in the text below.

Add all these lost lands to the current Swiss map and the result is overleaf: a new Greater Switzerland with 40 cantons instead of 26 (we made Constance the capital of Thurgau rather than a separate canton). This is a Switzerland that never was but might easily have been.

1. Unter Calven. In 1367, together with Val Müstair, it came under the control of the Three Leagues (later known as Graubünden). While Val Müstair became Swiss, Unter Calven became Austrian in 1499 and eventually Italian.

2. Valtellina or Veltlin. A large area controlled by the Three Leagues from 1512 to 1797, when it was forcibly transferred to the Cisalpine Republic. At the Congress of Vienna in 1815 it was declared Austrian and is now Italian.

3. Tre Pievi or Drei Pleven. Three communities, with only 21 villages, beside Lake Como. Controlled by the Three Leagues in 1512 but by 1531 they were Milanese.

4. Luino. The land between Lakes Lugano and Maggiore, including Luino, Cuvio and Valtravaglia was briefly Swiss from 1513 to 1515.

5. Val d'Ossola or Eschental. An obvious target for annexation as it sticks up into Switzerland like a Matterhorn-shaped stalagmite. The Swiss held it from 1410-22 and again in 1512-15 but lost it after their defeat at Marignano.

6. Chablais. The southern shore of Lake Geneva was taken after victory over Savoy in 1536. Bern had the part around Thonon, Valais had the area around Evian. By 1569 both had returned their spoils to Savoy.

▶

1. Unter Calven

2. Valtellina

3. Tre Pievi

4. Luino

5. Val d'Ossola

6. Chablais

7. Faucigny

8. Pays De Gex

9. Franche-Comté

10. Besançon

11. Montbéliard

12. Mulhouse

13. Rottweil

14. Constance

15. Vorarlberg

◀

1. Faucigny. In 1860 a petition advocating that Faucigny become a Swiss canton was signed by 7,606 people in 60 communities. But in April of that year a popular vote decided in favour of joining France instead.

2. Pays de Gex. A part of Savoy that Canton Bern snatched in 1536, only to lose it in 1564. One small bit became Swiss again in 1815 to create a bridge between Geneva and the rest of Switzerland.

3. Franche-Comté. Occupied by the Swiss during the Burgundian Wars in 1474-77, it was sold to the Habsburgs in 1478 for 150,000 guilders.

4. Besançon. The capital of France-Comté, Besançon was also separately a Swiss ally in the Middle Ages.

5. Montbéliard. Part of Württemberg (in present-day Germany) since the 15th century but also a Swiss ally and a centre of watch-making. French since 1815.

6. Mulhouse. The Alsatian city-state was allied first to Bern and Solothurn (1466), then to all the cantons in 1515. Its status as a Swiss ally remained until 1798.

7. Rottweil. In 1463 the city-state signed a pact with the Swiss, which became permanent in 1519 when Rottweil was given official allied status. That remained in force until 1697.

8. Constance. The now-German city almost became Swiss in the mid-15th century but rural cantons feared another big city upsetting the balance between town and country. The city's powerful bishop opted against the Swiss and Constance joined the Swabian League. If it had become Swiss, it would have been the natural capital of Thurgau.

9. Vorarlberg. In a 1919 referendum this Austrian province voted by 82% to become a Swiss canton. That was overruled by the Allies, and not welcomed by many Swiss who feared the country becoming too dominated by German-speakers.

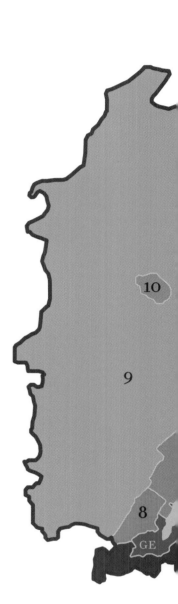

The individual maps appeared in 2014 on the Watson news website in an article about a possible alternative Switzerland. We added Constance and our designer transformed Watson's electronic animated map into a print version of Greater Switzerland.

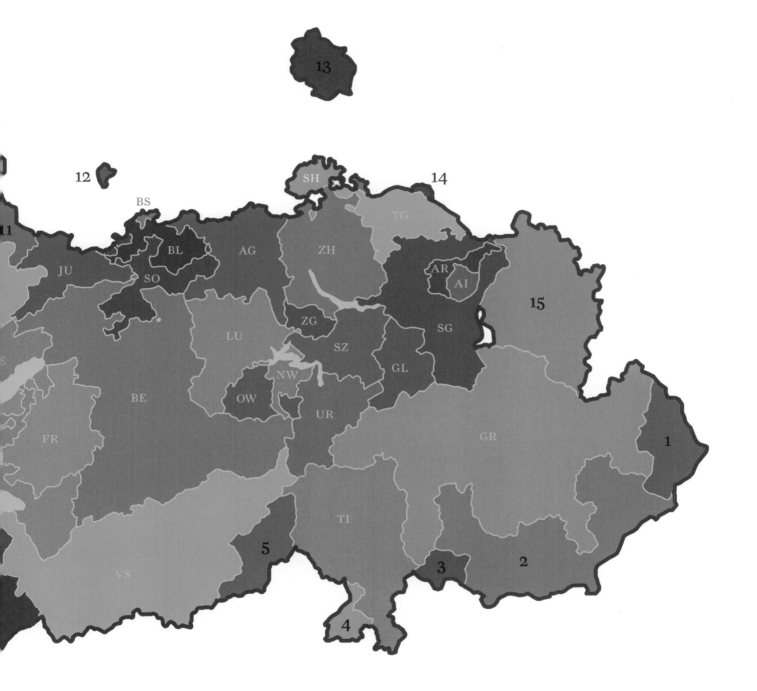

80

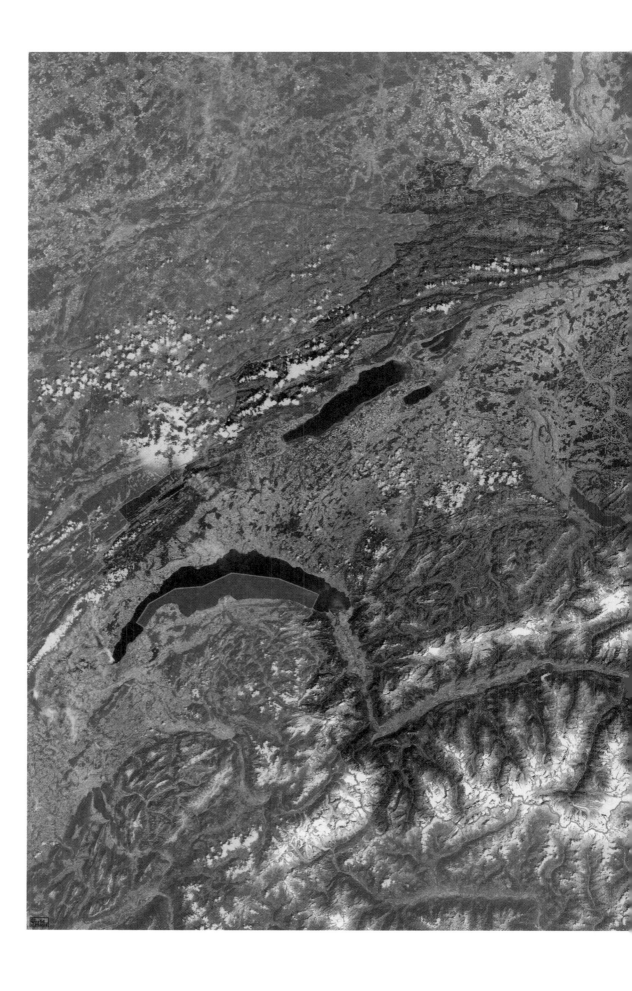

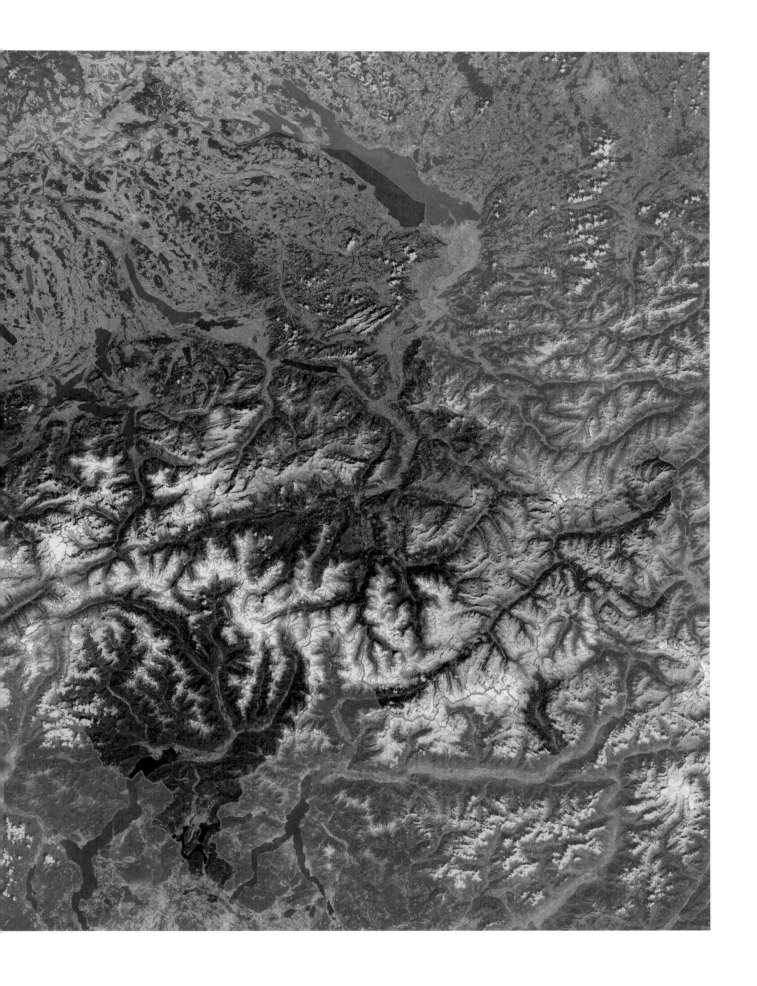

Appendix

The Swiss cantons

There are 26 cantons in Switzerland, all constitutionally equal with regard to almost every aspect of Swiss life. Except one. Politically, six of the 26 can sometimes be counted as half-cantons, even though some have larger populations than whole cantons. In terms of votes in a referendum and seats in the smaller house of the federal parliament, these six halves do not have the same status as the 20 wholes.

Here are all the cantons, listed in alphabetical order by their official two letter codes, along with the date of entry into the confederation, the capital and land area:

AG	**Aargau**	1803	Aarau	1,404 km²
AI	**Appenzell Innerrhoden**	1513[a]	Appenzell	173 km²
AR	**Appenzell Ausserrhoden**	1513[a]	Herisau	243 km²
BE	**Bern**	1353	Bern	5,959 km²
BL	**Basel-Landschaft**	1501[b]	Liestal	518 km²
BS	**Basel-Stadt**	1501[b]	Basel	37 km²
FR	**Fribourg**	1481	Fribourg	1,671 km²
GE	**Geneva**	1815	Geneva	282 km²
GL	**Glarus**	1352	Glarus	685 km²
GR	**Graubünden**	1803	Chur	7,105 km²
JU	**Jura**	1979	Delémont	838 km²
LU	**Lucerne**	1332	Lucerne	1,493 km²
NE	**Neuchâtel**	1815	Neuchâtel	803 km²
NW	**Nidwalden**	1291[c]	Stans	276 km²
OW	**Obwalden**	1291[c]	Sarnen	491 km²
SG	**St. Gallen**	1803	St. Gallen	2,026 km²
SH	**Schaffhausen**	1501	Schaffhausen	298 km²
SO	**Solothurn**	1481	Solothurn	791 km²
SZ	**Schwyz**	1291	Schwyz	908 km²
TG	**Thurgau**	1803	Frauenfeld	991 km²
TI	**Ticino**	1803	Bellinzona	2,812 km²
UR	**Uri**	1291	Altdorf	1,077 km²
VD	**Vaud**	1803	Lausanne	3,212 km²
VS	**Valais**	1815	Sion	5,224 km²
ZG	**Zug**	1352	Zug	239 km²
ZH	**Zurich**	1351	Zurich	1,729 km²

[a] The two half-cantons of Appenzell split in 1597
[b] The two half-cantons of Basel split in 1833
[c] The two half-cantons (together known as Unterwalden) have always been separate

Credits

All maps are north oriented unless otherwise stated. Places and dates of publication given where these are known; original sizes and scales given where this information was available.

Foreword: See map no. 7

Introduction: Alexis-Hubert Jaillot: *Les Suisses, leurs alliés et leurs sujets.* Paris 1703. 1 map over four sheets, altogether 130× 93 cm, ca. 1:230,000. Zentralbibliothek Zürich, 4 Hb 04: 1.

1: Albrecht von Bonstetten: *Karte der Eidgenossenschaft, Handschrift Superioris Germanie confoederationis descriptio.* Einsiedeln? 1480. Bibliothèque nationale de France, Paris, Ms. Lat. 5656, fol. 8, 22 × 15 cm.
From: **Claudius Sieber-Lehmann:** *Albrecht von Bonstettens geographische Darstellung der Schweiz von 1479.*
In: *Cartographica Helvetica 16* (1997), S. 39–46.

2: Konrad Türst: *De situ confoederatorum descriptio.* 1495/97? Manuscript map, 53 × 39 cm, ca. 1:500,000, southeast oriented. Zentralbibliothek Zürich, Ms Z XI 307a.

3: Martin Waldseemüller: *Tabula nova heremi Helvetiorum.* Strasbourg 1513. 52 × 41 cm, ca. 1:400,000, south oriented. Zentralbibliothek Zürich, 5 Hb 02: 1.

4: Johannes Stumpf: *Die Rauracer/Basler Gelegenheit.* Zürich 1547/48. 29 × 18 cm, ca. 1:600,000, south oriented. Zentralbibliothek Zürich, Ms A 18, Bl. 171r.

5: Aegidius Tschudi: *Helvetiae descriptio.* In: Abraham Ortelius: *Theatrum Orbis Terrarum.* Antwerpen 1570/71. 44 × 34 cm, ca. 1:900,000, south-southeast oriented. Zentralbibliothek Zürich, 3 Hb 02: 34.

6 & 14: Gerardus Mercator: *Helvetia cum finitimis regionibus confoederatis.* Duisburg 1595. 45 × 33 cm, ca. 1:800,000. Zentralbibliothek Zürich, 4 Hb 02: 12.

7: Heinrich Ludwig Muoss: *Helvetia, Rhaetia, Valesia. Das Schweitzerland, ein von Gott gesegneter Freyheit- und Fridenssitz und der Mit-Verpündten Vatter-Land.* Zug 1710. 73 × 52 cm, ca. 1:580,000, whole sheet 101 × 87 cm. Zentralbibliothek Zürich, 5 Hb 04: 15.

8: Johann Jakob Scheuchzer: *Nova Helvetiae tabula geographica illustrissimis et potentissimis cantonibus et respublicis reformatae religionis Tigurinae, Bernensi, Glaronensi, Basiliensi, Scaphusianae, Abbatis Cellanae.* Amsterdam 1715. 1 map with four sheets, each 55 × 45 cm, together 109 × 87 cm, ca. 1:320,000. Zentralbibliothek Zürich, 5 Hb 04: 11.

9: Johann Jakob Scheuermann: *Helvetische Republik Eingetheilt in Cantone und Districte.* Zürich 1799. 27 × 20 cm, ca. 1:1,300,000. Zentralbibliothek Zürich, 3 Hb 54: 2 Expl 2.

10: Robert Wilkinson, John Froggett: *Switzerland.* London 1808. 30 × 21 cm, ca. 1:1,200,000. Zentralbibliothek Zürich, 3 Hb 05: 69.

11: Anonymous: *Karte von der Schweiz.* Zürich 1815. 63 × 45 cm, ca. 1:500,000. Zentralbibliothek Zürich, 4 Hb 05: 17.

12: Eidgenössisches Topographisches Bureau: *Topographische Karte der Schweiz, Ausschnitt.* Vermessen und herausgegeben auf Befehl der eidgenössischen Behörden, aufgenommen unter Aufsicht des Generals Guillaume Henri Dufour.
Upper map: Geneva 1862. 70 × 48 cm, 1:100,000. Zentralbibliothek Zürich, Kupfer 95: 23 Ed. 1862.
Lower map: Bern 1926. 70 × 48 cm, 1:100,000. Zentralbibliothek Zürich, 5 Hb 75: 2: 23 Ed. 1926

13: Jean Duvillard: *Carte du Léman.* 1588. Fische von 1581. 102 × 38 cm, south oriented. Bibliothèque de Genève, Centre d'iconographie genevoise, CIG ms. fr. 140.

14: see map 6

15: Gabriel Walser: *Das Land Appenzell der Innern u. Aussern Rooden.* St. Gallen 1740. 36 × 20 cm, ca. 1:140,000, south oriented. Zentralbibliothek Zürich, 3 Jb 04: 1.

16: Martin Martini: *Grosse Freiburger Stadtansicht.* Freiburg 1606. Copper engraving on 8 plates, 156 × 86 cm, View from the South, Important Buildings emphasised with a different Scale. Musée d'Art et d'Histoire Fribourg.

17: Antoine Lambien: No Title. Lyon 1709. 62 × 42 cm, 1:350,000. Mediathèque Valais-Sion, Special Collections: KF 8.

18: Johann Heinrich Streulin: *Zürich Gebiet.* Zürich? 1698, 26 × 28 cm, ca. 1:250,000, northeast oriented. Zentralbibliothek Zürich, 31 Kb 03: 1.

19: Jakob Störcklein: *Nova ditionis Bernensis Tabula Geographica Ursi effigie delineata.* Basel 1700. 33 × 23 cm, ca. 1:800,000. Universitätsbibliothek Bern, Ryh 3211: 25b.

20: Johann Jakob Scheuchzer: *Rheni, Rhodani, Ticini, Ursae, prima stamina in sumis alpibus helveticis.* Leiden, 1723. 45 × 27 cm, ca. 1:115,000. Zentralbibliothek Zürich, 3 Hf 04: 1.

21: Jean Rocque: *A plan of Geneva and the Environs.* London? 1760. 62 × 47 cm, ca. 1:5,000, northeast oriented. Bibliothèque de Genève, Centre d'iconographie genevoise.

22: Georg Christoph Kilian: *Neufchastel, oder souveraines Fürstenthum Neüenburg in Schweizerischen Bund, nebst der Graffschaft Vallangin.* Augsburg 1757–1780. 21 × 16 cm, ca. 1:440,000. Zentralbibliothek Zürich, 3 Jk 04: 7.

23: Anonymous: *Plan perspectif d'une grande partie des Cantons de Lucerne, d'Uri, de Schwyz, d'Unterwalden, de Zoug, et de Glaris, avec la frontière de celui de Berne.* 1786. 61 × 26 cm, southeast oriented. Zentralbibliothek Zürich, 4 Hl 04: 5.

24: J. Lalive: *Plan de La Chaux de Fonds. Avec les alignements et nivellements fixés.* La Chaux-de-Fonds 1887. 72 × 52 cm, 1:3,000, northwest oriented. Zentralbibliothek Zürich, 4 Jk 06: 3.

25: Heinrich Keller: *Der Canton Ticino.* Zürich 1812. 20 × 26 cm, ca. 1:350,000, Nullmeridian Ferro. Zentralbibliothek Zürich, 3 Jp 05: 2.

26: Johann Zuber: *Grundriss der Stadt St. Gallen, nebst der nächsten Umgebung trigonometrisch aufgenommen von Joh. Zuber.* St. Gallen 1828. 74 × 53 cm, ca. 1:2,880, southeast oriented. Copyright: Vermessungsamt der Stadt St. Gallen.

27: Eduard Beck: *Neuster Plan der Stadt Bern und den Umgebungen.* Bern 1861. 67 × 49 cm, 1:4 167. Zentralbibliothek Zürich, 4 Jd 06: 12.

28: Joseph Krottendorfer: *Uibergang der Franzosen über die Limmat und Schlacht bey Zürich am 25ten September 1799.* Ohne Ort, after 1800. 51 × 34 cm, ca. 1:30,000, south-southwest oriented. Zentralbibliothek Zürich, 4 Kw 55: 2.

29: Franz Malté: *Der Kriegsschauplatz in der Schweiz.* Augsburg or München 1847. 22 × 14 cm, ca. 1:1,725,000, Zentralbibliothek Zürich, 3 Hb 55: 13.

30: Anonymous: *Lageplan der Völkerbund-Gebäude im Arianapark, Geneva.* Schweizerische Bauzeitung, 5. Dezember 1931. 15 x 13 cm, 1:5,000, west-northwest oriented. Zentralbibliothek Zürich, UU 109bc, Bd. 98, Nr. 23, S. 291 (5. Dezember 1931).

31: Deutsche Wehrmacht, Oberkommando des Heeres: *Operation Tannenbaum.* H.Gru. Kdo. C Ia Nr. 262/40 g.K.v. 4.10.1940. 115 x 75 cm, 1:300,000. 1 Bundesarchiv, Abteilung Militärarchiv, Freiburg im Breisgau, BA-MA, RH 2/465, K-10.

32: Schweizer Armee, Armeekommando, Gruppe Id, Kartographische Abteilung: *Internierungslager, Lage am 16. 12. 40.* Bern 1940. Section from one of 9 maps, each 117 × 74 cm, 1:300,000. ETH-Bibliothek Zürich, KS K 620103: 6 (16.12.1940), reproduced with the permission of swisstopo (BA140271).

33: Schweiz. Eidgenössisches Statistisches Amt: *Verbreitung der Schweinerassen in der Schweiz, nach den Ergebnissen der eidg. Viehzählung vom 21. April 1941.* Bern 1944. 57 × 39 cm, 1:600,000. Zentralbibliothek Zürich, 4 Hb 47: 13.

34: Armeekommando, Operationssektion: *Schweizer Armee, Operationsbefehl 13, 24. Mai 1941.* Bern 1941. 1:1,000,000. Schweizerisches Bundesarchiv Bern, CH-BAR E-27 1000/721 14299.

35: British Directorate of Military Intelligence MI9: *Escape map on silk, section from the whole map with France, Germany, Switzerland, Croatia, Montenegro, Hungary, Slovakia, Italy;* Bl. 87 × 74 cm, gef. 10 × 11 cm, 1:1,000,000. Zentralbibliothek Zürich, 28 CA 87: 1: 5

36: SSSR. General'nyi Štab: *Bazel', Muttenc, Riën, Vejl'-am-Rejn, Lërrach, Sen-Lui.* Moscow 1975. 2 maps, each 115 × 56 cm, with street index, 1:10,000. Zentralbibliothek Zürich, 5 Jc 08: 3 1+2.

37: Franz Johann Joseph von Reilly: *Postkarte von der Schweiz = Repraesentatio cursuum publicorum in omnibus Helvetiae partibus.* Wien 1799. 43 × 31 cm, ca. 1:860,000, Nullmeridian Ferro? Zentralbibliothek Zürich, 4 Hb 44: 3.

38: Aristide Michel Perrot: *Route du Simplon.* Paris 1824. 42 × 30 cm, ca. 1:100,000. Mediathèque Valais-Sion, Special Collections, KD Sim1: 01.

39: Robert Stephenson, Henry Swinburne, Auguste d'Ivernois: *Croquis pour l'intelligence du rapport du 12 octobre 1850 sur l'établissement de chemins de fer en Suisse, (feuille fédérale no. 52).* Geneva 1851. 48 × 33 cm, 1:926,500. Zentralbibliothek Zürich, 3 Hb 46: 6.

40: Rudolf Gross: *Karte und Panorama vom Rigi.* Zürich 1860? Map diameter 24 cm, Bl. 38 × 37 cm, gef. 12 × 22 cm, ca. 1:81,000. Zentralbibliothek Zürich, 3 Jn 06: 7.

41: Thomas Cook: *Cook's Railway Map of Switzerland, Savoy, the Italian Lakes & c.* London 1873. 24 × 17 cm, 1:600,000. From: *Cook's Continental Time Tables & Tourist's Handbook.* Thomas Cook Archives, Peterborough (England): 13.196.

42: Anonymous: *Gotthard-Bahn, Laghi di Como, Maggiore & Lugano.* Mailand 1902. 74 × 127 cm. Museum für Gestaltung Zürich MfGZ, Plakatsammlung: 26-0359.

43: Anonymous: *Berne–Loetschberg–Simplon.* Bern 1920. Copyright: BLS-AG, Bern.

44: Statistisches Bureau des eidgenössischen Departements des Innern: *Die historische Entwicklung des schweiz. Eisenbahnnetzes.* Bern 1914. 40 × 27 cm, scale: 1:1,000,000. Bibliothek des Bundesamtes für Statistik, Neuchâtel, BFS 10100 01 11 325011 1100.

45: Schweizerisches Post- und Eisenbahndepartement: *Höhenprofil-Karte der schweizerischen Eisenbahnen.* Bern 1915. 46 x 37 cm, length = 1:250,000, height = 1:10,000, horizon = sea level. Bibliothek des Bundesamtes für Statistik, Neuchâtel, BFS 10100 71 14 000000 0001 0.

46: Swissair: *Swissair routes.* Bern 1949. 55 × 44 cm, gef. 13 × 24 cm, 1:32,000,000, north-northwest oriented. Zentralbibliothek Zürich, 16 Ba:47: 3, Copyright: Hallwag Kümmerly+Frey AG Schönbühl.

47, 48, 49: Schweizerische Kommission für die Planung des Hauptstrassennetzes: *Das schweizerische Nationalstrassennetz.* Bern 1958. From: *Zusammenfassender Bericht der Kommission des Eidg. Departementes des Innern für die Planung des Hauptstrassennetzes.* Gesamthaft 85 Seiten, 37 Blätter, Tafeln, Karten, Tabellen. Zentralbibliothek Zürich, Kart 10042: Fig. 33 (Whole of Switzerland), 21 (Zurich), 27 (Geneva).

50: Otto M. Müller: *Die Bernina-Linie der Rhätischen Bahn.* Early 1960s. 26 × 18 cm, west-southwest oriented. Zentralbibliothek Zürich, MK 1179.

51: Anonymous: *Zürich. Stadtplan für Männer.* Berlin, 1970er-Jahre. 47 × 38 cm, gef. 13 × 21 cm, ca. 1: 8,600, east oriented. Inset map: «Vergnügungsviertel Niederdorf». Reverse side with lists of bars. Zentralbibliothek Zürich, 16 Lb 38: 2. Monika Dülk Verlag, Berlin.

52: Transports publics de la région lausannoise SA: *Plan du réseau jusqu'au 1er juin 1991.* Ohne Ort, 1989. Copyright: Transports publics lausannois tl.

53: Winfried Kettler: *Der Weg der Schweiz.* 1989. 38 × 58 cm, south oriented. Zentralbibliothek Zürich, 4 Js 48: 1.

54: Anonymous: No title, 2014. Southeast oriented. Copyright: Jungfraubahnen 2014, Interlaken.

55: Heinrich C. Berann: Postcard with no title, 1910. Southeast oriented. Jungfraubahnen 2014, Interlaken.

56: Ursula Hitz: *Schweizer Berge.* London 2014. Copyright: Ursula Hitz, www.ursulahitz.com.

57: AlpTransit Gotthard AG: 3 plans from: *AlpTransit Gotthard – Neue Verkehrswege durch das Herz der Schweiz,* Nr. 12, Lucerne 2011. Page 5: Flachbahn. Page 13: Übersicht. Page 18: Längenprofil. Copyright: AlpTransit Gotthard AG.

58: Hans Wilhelm Auer: *Parlamentsgebäude Schweizerisches Bundesparlament, Bern, 1. Stock, Grundriss.* Bern 1901. Scale: 52 x 44 cm, 1:200, south oriented. Bundesamt für Bauten und Logistik BBL, Bern.

59: Statistisches Bureau des eidgenössischen Departements des Innern: *Die Sterbefälle infolge von Selbstmord im Verhältnis zur Bevölkerung, nach den Durchschnittsergebnissen der Jahre 1901–1910.* Bern 1914. 40 × 27 cm, scale: 1:1,000,000. Bibliothek des Bundesamtes für Statistik, Neuchâtel, BFS 10100 01 11 325011 1100.

60: First map: Statistisches Bureau des eidgenössischen Departements des Innern: *Die Häufigkeit der Eheschliessungen im Verhältnis zur Wohnbevölkerung, nach dem Durchschnittsergebnis der Jahre 1901–1910.* Bern 1914.
Second map: **Bundesamt für Statistik:** *Rohe Heiratsziffer, 2001–2010.* Neuchâtel 2014.
Both 40 × 27 cm, scale: 1:1,000,000.
Bibliothek des Bundesamtes für Statistik, Neuchâtel.

61: First map: Statistisches Bureau des eidgenössischen Departements des Innern: *Die Häufigkeit der Geburten im Verhältnis zur Bevölkerung, nach dem Durchschnittsergebnis der Jahre 1901–1910.* Bern 1914.
Second map: Bundesamt für Statistik: *Rohe Geburtenziffer, 2001–2010.* Neuchâtel 2014. Both 40 × 27 cm, scale: 1:1,000,000.
Bibliothek des Bundesamtes für Statistik, Neuchâtel.

62: First map: Statistisches Bureau des eidgenössischen Departements des Innern: *Die Rindviehhaltung in der Schweiz nach den Ergebnissen der eidg. Viehzählung vom 21. April 1911.* Bern 1914.
Second map: **Bundesamt für Statistik:** *Die Rindviehhaltung in der Schweiz, 2011.* Neuchâtel 2014. Both 40 × 27 cm, scale: 1:1,000,000. Bibliothek des Bundesamtes für Statistik, Neuchâtel.

63: Carl Künzli: *Souvenir de la Suisse.* Zürich 1914. 14 × 9 cm, private archives Diccon Bewes, Bern, Editions Cartes Postales Kunzli Zurich (Dep. No. 2128).

64: Otto M. Müller: *Käsekarte der Schweiz.* Bern 1975. 43 × 30 cm, ca. 1:1,000,000. Published by Schweizerischen Käseunion. Bern 1975. Zentralbibliothek Zürich, 3 Hb 48: 5.

65: Tages-Anzeiger: *Beitritt der Schweiz zum EWR.* Zürich, 7. Dezember 1992 from an article „Europa einigt sich – Schweiz gespalten" on the front page. Zentralbibliothek Zürich, UZ 31, 100. Jg., Nr. 285, Seite 1 (7.12.1992).

66: Patrick Chappatte: No title. Geneva, 1990s. Copyright: www.globecartoon.com.

67: Bundesamt für Statistik: *Landessprachen in den Gemeinden, 2000.* Neuchatel 2014. 29,7 × 21 cm, scale: 1:1,500,000. Neuchâtel. Bundesamt für Statistik.

68: Michael Hermann, Heiri Leuthold: *Politische Landkarte der Schweiz. From: Atlas der politischen Landschaften. Ein weltanschauliches Porträt der Schweiz.* Vdf Hochschulverlag an der ETH Zürich 2003. Copyright: Forschungsstelle Sotomo, Michael Hermann, Heiri Leuthold.

69: Tourisme Neuchâtelois: *Watch Valley Jura & 3 Lacs.* Neuchâtel 2004. 63 × 42 cm, ca. 1:71,000, northwest oriented. Copyright: Tourisme Neuchâtelois, Neuchâtel.

70: atheist raskalnikow: *Mittlere ständige Wohnbevökerung, 2011.* März 2013. Copyright: raskalnikow.wordpress.com, Blogpost Nr. 132, 5. März 2013.

71: Club School Migros: *Regionale Genossenschaften der Migros-Gruppe.* Zürich 2014. Copyright: Klubschule Migros, Zürich.

72, 73: Pieter van den Keere: *Représentation du plan et assiette de la novelle ville nommée Henripolis qui se bastit proche de Neufchastel en Suisse.* Werbeschrift, 1626. 41 × 32 cm, northwest oriented. Supplement in a brochure. Zentralbibliothek Zürich, 3 Jk 03:3.

74: Wurster, Randegger & Co.: *Passage du Lucmanier, plan général.* Winterthur 1860? 76 × 47 cm, 1:100,000, west-northwest oriented. Zentralbibliothek Zürich, 4 Hf 46:8.

75: Walter Minder: *Eine Autostrasse über den Gemmipass von Kandersteg nach Leukerbad.* Bern 1952.
Illustration: *Die Gemmistrasse im Üschinental mit Alpbachviadukt und Üschinengrattunnel.* Federzeichnung von Etienne Clare.
Map: *Situationsplan der Gemmistrasse,* ca. 25 × 15 cm, 1:80,000. Published by Initiativkomitee für den Bau einer Autobahn über die Gemmi. Zentralbibliothek Zürich, DU 2907. Copyright: Etienne Clare Erbengemeinschaft.

76: Anonym: *U-Bahn, Region Zürich; 1. Etappe: Kloten– Hauptbahnhof-Dietikon.* Zürich 1973. 18 × 12 cm, ca. 1:200,000. Zentralbibliothek Zürich, 3 Kc 48:2.

77: Pierre-Alain Rumley: *Switzerland with 13 cantons.* La Chaux-de-Fonds 2010. From: La Suisse demain – utopie ou réalité? Copyright: Pierre-Alain Rumley, Couvet.

78: Pro Swissmetro: *Swissmetro 2014 – der unterirdische Inter-City.* www.swissmetro.ch 2014. The website also shows a cross-section of the main tunnels and service tunnels. Copyright: www.swissmetro.ch 2014, Stand 17. 3. 2014.

79: Adapted presentation from Watson original: „*Die Schweiz, das kleine Russland – so gross könnte die Eidgenossenschaft wirklich sein.*" www.watson.ch 2014. 16 maps adapted by Cécile Gretsch (original: www.watson.ch, 28. 3. 2014)

80: Landsat 7: *Switzerland from space.* Ottawa 2001. World of Maps: satellite picture 2001. Copyright: www.worldofmaps.com.

Thanks

Switzerland is not an island, at least not physically, and neither is any book solely the result of one person's work. That this book exists at all is down to Madlaina Bundi at Hier+Jetzt Verlag, who believed in it from the outset and accompanied me through all the stages of creation and production. It's been quite a cartographical journey for both of us, and we were helped along the way by everyone at Hier+Jetzt.

Just as important to the book's birth was Markus Oehrli, the archivist at the Zurich Central Library, who guided me through the library's extensive map collection even when I didn't really know what I was looking for. His ideas and insight helped the book grow into what it is now. The maps played a role too.

Thomas Schulz at the Federal Statistics Office in Neuchâtel was an invaluable source of information, both in tabular and cartographic form, as well as an enthusiastic supporter of this project. He also very kindly agreed to write the Foreword in his role as President of the Swiss Society of Cartography. Thanks to Ian White for introducing us to each other.

Thank you also to all the others who helped me with my map quest, especially, Christine Falcombello and Sarah Chapalay at Centre d'iconographie genevoise, Paul Smith at Thomas Cook, Daniel Huber at watson.ch, Thomas Klöti at Bern University, Pierre-Alain Rumley at Neuchâtel University, Céline Epars at Lausanne Transport, Simon Roth at Mediathek Wallis, Martina Stercken at Zurich University, Peter Barber at the British Library, Ursula Hitz, and of course Patrick Chappatte for letting us use his cartoon.

Not forgetting Hadi Barkat and everyone at Helvetiq, who made both the English and French editions of this book possible.

To my agent, Alex Christofi, thank you for waiting patiently for me to finish this labour of love so that we can then work together on future ideas, some of which may not involve Switzerland at all.

Lastly, as always, thanks to my family and Gregor for the love and support that make all this possible. ❤

Impressum

This second edition (paperback) published by Helvetiq
in 2017

ISBN 978-2-940481-30-9

Graphic Design: Renate Zeller, Laura Tobler

Legal deposit in Switzerland
October 2015
Bibliothèque Cantonale, Lausanne, Vaud / Waadt
Schweizerische Nationalbibliothek, Berne / Bern

UK Distributor: Turnaround (www.turnaround-uk.com)

© 2015, Helvetiq

HELVETIQ

Côtes de Montbenon 30
1003 Lausanne, Switzerland

Vogesenplatz 1
4056 Basel, Switzerland

info@helvetiq.ch
www.helvetiq.ch